HEALTH EFFECTS INVESTIGATION OF OIL SHALE DEVELOPMENT

HEALTH EFFECTS INVESTIGATION OF OIL SHALE DEVELOPMENT

Edited by

W. H. Griest

M. R. Guerin

Analytical Chemistry Division
Oak Ridge National Laboratory
Oak Ridge, Tennessee

D. L. Coffin

Health Effects Research Laboratory
U.S. Environmental Protection Agency
Research Triangle Park, North Carolina

Proceedings of the Symposium on Health Effects Investigation of Oil Shale Development

Gatlinburg, Tennessee
June 23–24, 1980

Sponsored by
U.S. Environmental Protection Agency

ANN ARBOR SCIENCE
PUBLISHERS INC / THE BUTTERWORTH GROUP

230 Collingwood, P.O. Box 1425, Ann Arbor, Michigan 48106

Library of Congress Catalogue Card No. 80-70323
ISBN 0-250-40169-X

Manufactured in the United States of America

SYMPOSIUM COMMITTEE

W. H. Griest
M. R. Guerin
D. L. Coffin

K. M. Branam and P. J. Mullins
Conference and Proceedings Secretaries

FOREWORD

Environmental and health issues must be addressed as an integral part of synfuels technology development. To this end, the U.S. Environmental Protection Agency, the Department of Energy, Department of Defense (U.S. Navy), the American Petroleum Institute, Development Engineering, Inc., and the Standard Oil Company are cooperating in an interagency matrix approach investigation of the potential health effects associated with the production and refining of shale oil into jet and diesel fuels. Groups of biologists, ecologists, and chemists in laboratories across the nation are involved in this multidisciplinary research effort. This volume is a record of the findings to date as presented at the symposium, Health Effects Investigation of Oil Shale Development, held at Gatlinburg, Tennessee, June 23-24, 1980, sponsored by the U.S. Environmental Protection Agency under Interagency Agreement DOE No. 40-601-76, EPA No. 79-D-X0319-1, under Union Carbide Corporation contract W-7405-eng-26 with the U.S. Department of Energy. The progress summarized here illustrates the success of the "matrix" approach and suggests that other materials and technologies can be similarly studied.

D. L. Coffin
U.S. Environmental Protection Agency
Research Triangle Park, North Carolina

W. H. Griest leads the Quantitative Analysis Group of the Bio/Organic Analysis Section in the Analytical Chemistry Division of Oak Ridge National Laboratory. His analytical methods developments and applications to environmental, occupational, and industrial samples have led to more than 15 publications. He holds a BS in chemistry from Virginia Polytechnic Institute and State University, and a PhD in physical chemistry from Vanderbilt University.

M. R. Guerin holds a BS in chemistry from Northern Illinois University and a PhD in analytical chemistry from Iowa State University. He is currently head of the Bio/Organic Analysis Section in the Analytical Chemistry Division of the Oak Ridge National Laboratory. His research activities in the areas of organic analytical chemistry and bioassay chemistry have resulted in approximately 50 publications.

D. L. Coffin is a Senior Science Advisor in the U.S. Environmental Protection Agency Health Effects Research Laboratory at Research Triangle Park, North Carolina. He holds a VMD from the University of Pennsylvania School of Veterinary Medicine. His research activities in veterinary pathology and environmental and health problems of synfuels and mineral fibers have resulted in more than 100 publications.

CONTENTS

LIST OF CONTRIBUTORS

A. Austin, U. S. Environmental Protection Agency, Environmental Research Center, Research Triangle Park, NC 27711

S. C. Blum, Exxon Research and Engineering, Linden, NJ 07036

R. Bradow, U.S. Environmental Protection Agency, Environmental Research Center, Research Triangle Park, NC 27711

A. Carrano, Lawrence Livermore Laboratory, Biomedical Sciences Division, Livermore, CA 94550

J. Carver, Battelle Columbus Laboratories, Biomedical Sciences Division, Columbus, OH 43201

D. L. Cawein*, Director, Engineering, Health, Environmental Affairs, Standard Oil Company, Cleveland, OH 44115

B. R. Clark, Analytical Chemistry Division, Oak Ridge National Laboratory, P.O. Box X, Oak Ridge, TN 37830

L. Claxton, U.S. Environmental Protection Agency, Environmental Research Center, Research Triangle Park, NC 27711

G. M. Clemmer, Biology Division, Oak Ridge National Laboratory, P.O. Box X, Oak Ridge, TN 37830

D. L. Coffin*, Health Effects Research Laboratory, U.S. Environmental Protection Agency, MD-51, Research Triangle Park, NC 27711

LCDR M. J. Cowan*, MSC, USN, Naval Medical Research Institute Toxicology Detachment (NMRI/TD), Wright-Patterson AFB, OH 45433

J. L. Epler, Biology Division, Oak Ridge National Laboratory, P.O. Box X, Oak Ridge, TN 37830

J. S. Felton, Lawrence Livermore Labortory, Biomedical Sciences Division, Livermore, CA 94550

M. E. Foreman, Toxicology Group, Life Sciences Division, Los Alamos National Scientific Laboratory, Los Alamos, NM 87544

---

*Speakers

J. M. Giddings*, Environmental Sciences Division, Oak Ridge National Laboratory, P.O. Box X, Oak Ridge, TN 37830

L. C. Gipson, Biology Division, Oak Ridge National Laboratory, P.O. Box X, Oak Ridge, TN 37830

W. H. Griest*, Analytical Chemistry Division, Oak Ridge National Laboratory, P.O. Box X, Oak Ridge, TN 37830

M. R. Guerin, Analytical Chemistry Division, Oak Ridge National Laboratory, P.O. Box X, Oak Ridge, TN 37830

W. M. Haschek, Biology Division, Oak Ridge National Laboratory, P.O. Box X, Oak Ridge, TN 37830

F. T. Hatch, Lawrence Livermore Laboratory, Biomedical Sciences Division, Livermore, CA 94550

R. N. Heistand*, Development Engineering Inc., Paraho Development Corporation, Grand Junction, CO 81501

J. M. Holland*, Biology Division, Oak Ridge National Laboratory, P.O. Box X, Oak Ridge, TN 37830

L. M. Holland*, Toxicology Group, Life Sciences Division, Los Alamos National Scientific Laboratory, Los Alamos, NM 87544

J. L. Huisingh, U.S. Environmental Protection Agency, Environmental Research Center, Research Triangle Park, NC 27711

CDR L. J. Jenkins, Jr., MSC, USN, Naval Medical Research Institute Toxicology Detachment (NMRI/TD), Wright-Patterson AFB, OH 45433

R. H. Jungers, U.S. Environmental Protection Agency, Environmental Research Center, Research Triangle Park, NC 27711

M. J. Kland*, Energy and Environmental Division, Lawrence Berkeley Laboratory, Berkeley, CA 94720

S. C. Lewis*, Exxon Corporation, P.O. Box 235, Mettlers Road, East Millstone, NJ 08873

T. K. Rao*, Biology Division, Oak Ridge National Laboratory, P.O. Box X, Oak Ridge, TN 37830

W. K. Robbins*, Exxon Research and Engineering, Linden, NJ 07036

L. H. Smith*, Biology Division, Oak Ridge National Laboratory, P.O. Box X, Oak Ridge, TN 37830

C. W. Stallard, Jr.*, M.D., The Standard Oil Company, Cleveland, OH 44115

T. J. Stephens, Biology Division, Oak Ridge National Laboratory, P.O. Box X, Oak Ridge, TN 37830

D. S. Stuermer, Lawrence Livermore Laboratory, Biomedical Sciences Division, Livermore, CA 94550

J. Sturm, U.S. Department of Transportation, Transportation Systems Center, Cambridge, MA 02142

---

*Speakers

L. H. Thompson, Lawrence Livermore Laboratory, Biomedical Sciences Division, Livermore, CA 94550

H. Timourian*, Lawrence Livermore Laboratory, Biomedical Sciences Division, Livermore, CA 94550

R. Walter, U.S. Department of Transportation, Transportation Systems Center, Cambridge, MA 02142

M. J. Whitaker, Biology Division, Oak Ridge National Laboratory, P.O. Box X, Oak Ridge, TN 37830

J. S. Wilson, Toxicology Group, Life Sciences Division, Los Alamos National Scientific Laboratory, Los Alamos, NM 87544

H. Witschi, Biology Division, Oak Ridge National Laboratory, P.O. Box X, Oak Ridge, TN 37830

D. A. Wolf, Computer Sciences Division, Union Carbide Corporation, Oak Ridge, TN 37830

L. B. Yeatts, Jr., Analytical Chemistry Division, Oak Ridge National Laboratory, P.O. Box X, Oak Ridge, TN 37830

S. Zimmering*, Division of Biology and Medicine, Brown University, Providence, RI 02912

R. Zweidinger, U.S. Environmental Protection Agency, Environmental Research Center, Research Triangle Park, NC 27711

---

*Speakers

CHAPTER 1

OVERVIEW OF THE REPOSITORY PROGRAM AND THE MATRIX APPROACH TO THE HEALTH EFFECTS INVESTIGATION OF OIL SHALE DEVELOPMENT

D. L. Coffin[a], M. R. Guerin[b], and W. H. Griest[b], Health Effects Research Laboratory[a], U.S. Environmental Protection Agency, Research Triangle Park, North Carolina 27711, and Analytical Chemistry Division[b], Oak Ridge National Laboratory, Oak Ridge, Tennessee 37830

INTRODUCTION

The production of petroleum substitutes from coal and oil shale is a crucial component of national efforts to achieve energy independence. It is generally agreed that environmental and health issues should be addressed in concert with engineering issues. Unexpected ecological and biological impacts might thus be identified early enough in the process development sequence to be avoided or eliminated when commercialization occurs. Further, synfuels technologies are certain to involve materials of unique chemical and physical properties as compared to petroleum processing. It is therefore expected that products and by-products will be the subject of intensive chemical and biological testing requirements by federal regulatory agencies. Early studies using synfuels related samples are essential in order to develop and validate reliable chemical and biological testing methods.

The purpose of this program is to develop suitable methods and to provide timely information on the biological properties of important synfuels technology components. A secondary purpose is to evaluate the effectiveness of a repository/matrix approach to carrying out such studies. To this end, the U.S. Environmental Protection Agency and the U.S. Department of Energy are cooperating with the U.S. Department of Defense (Department of the Navy), Development Engineering, Inc., the Standard Oil Company of Ohio (SOHIO), and the American Petroleum Institute in a concerted study of shale-derived crude and refined petroleum substitutes originating from above ground retorting using the Paraho Process.

This investigation is unique in at least three ways. First, it enjoys the close cooperation of an especially diverse group of agencies, institutions, and perspectives. Second, it involves the study of samples form a large scale (100,000 bbl) production and refining experience. Third, it relies on mutual scientific interests and exchanges of materials and results for continuing progress rather than on programmatic funding and direction.

The Matrix/Repository Approach

The term "matrix" derives from a series of investigators studying a common series of materials. The investigators (or more importantly their research areas, test systems, data contributions, etc.) constitute one axis of the matrix and the samples studied constitute the other axis of the matrix. An example for the study described in this volume is mutagenicity, skin carcinogenicity, physical properties, polycyclic aromatic hydrocarbon content, etc., versus Crude Shale Oil, Hydrotreated Shale Oil, shale derived DFM, etc. The results of a diverse number of groups focusing on diverse issues are made interrelatable by the fact that they examine or use a common set of samples.

Sample versus result matricies are a commonly used bookkeeping format for extensive process-specific or site-specific assessment programs. Specific sampling points are identified, the chemical and/or biological data required to characterize the samples are selected, and the matrix is filled-in as the program proceeds. Experimental activities are carefully programmed and monitored to ensure that the proper data are generated on the proper sample in the proper sequence. The approach requires central funding and management and requires that test methods be available and applicable on a routine basis so that schedules may be met.

The matrix approach being tested here differs from that required for site- or process-specific assessment. Its intent is to maximize the utility of individual investigator's research findings to the synfuels industry without infringing on the investigator's research committments to his/her sponsoring agency or freedom to publish or otherwise disseminate results. This is achieved by making samples available to the investigators which are more valuable (either in terms of relevance to the synfuels industry or because complementary chemical or biological data are available on the sample) than can be obtained independently.

Continued interaction is then fostered by providing investigators and sample suppliers with updated information on the status of related research on the samples. The investigator axis of the matrix thus evolves to include any who express a serious interest in contributing to the program. In this way, researchers addressing fundamental issues or developing advanced methods are helped to contribute to the more applied programs without requiring redirection of their efforts. In exchange, sample suppliers are provided with data and findings from the "basic research" community which would otherwise be unavailable and without the normally attendant cost.

The Research Materials Facility

A matrix program of the type outlined above requires a source of samples and information exchange. This was established[1,2,3] at the Oak Ridge National Laboratory by way of an interagency agreement between the U.S. Environmental Protection Agency (Health Effects Research Laboratory, Research Triangle Park) and the U.S. Department of Energy (Office of Health and Environmental Research). Responsibilities of the facility (colloquially, the "repository") staff initially included storing, chemically characterizing, and distributing samples as required by the USEPA or USDOE project officers. Responsibilities have expanded to include an active role in recruiting participants for matrix studies, disseminating research results to sample suppliers, and providing special services in support of sampling or high-priority research efforts identified by the project officers.

The Facility currently houses several hundred samples related to coal combustion, coal liquefaction, and oil shale retorting. Petroleum crudes and some refined products are also available. In addition, special arrangements are occasionally made with process developers for bulk quantities of materials for single study by others. Most of these materials receive little attention, however, as compared to those which are chosen for matrix studies.

A listing[2] of the current inventory of samples is available. Requests for samples are processed by informal contact with the sample suppliers. The plans of the requesting group are communicated to the supplier and guidance as to whether more suitable samples are available is sought. Samples and information concerning their previous and current study by others is forwarded to the requesting investigator following verbal approval of the supplier.

The amount of attention given to the samples depends on the intensity of their study. Infrequently requested samples are stored under refrigeration and protected from light.

Samples undergoing matrix studies are stored as above and, in addition, aliquots are stored under deepfreeze conditions in various containers. The bulk-stored and aliquot-stored samples are periodically screened for chemical stability. Samples undergoing extensive study by others are chosen for detailed chemical characterization as possible within existing resources.

A summary of "repository" activities is diagrammed in Figure 1. Emphasis is placed on the sample distribution, information exchange, and special services aspects of the program. It is not the intent of the activity to archive samples because they currently represent very early stages of process development and it is expected that the samples will change in nature with time. The objective is to extract the maximum amount of useful information possible and replace existing samples with samples of greater current relevance. A possible exception (currently under investigation) is the establishment of comparative research materials (CRMs). These materials, currently one coal oil, one shale oil, and one petroleum crude oil, are available in sufficient quantity for extensive interlaboratory comparison and detailed analysis to serve as common materials for validating methods and establishing fundamental chemical-biological relationships.

Comment

It must be emphasized that the samples available through the "repository" are not suitable for process- or site-specific research. Process conditions and sampling procedures at the time of sampling are not generally well enough specified. More importantly, sample handling procedures at the laboratories undertaking the research are seldom specified. It can be assumed that samples are carefully preserved by the investigators but it is not currently known whether any degree of care will ensure the integrity of the sample. The relevance of the results generated by the life scientists to the actual conditions at the commercial scale evolution of any given process is left to the sample suppliers and their process engineers. Methods developed as a result of these studies and fundamental relationships between the chemical composition and biological effects of the materials discovered through these studies are expected to be relevant and to contribute to environmentally acceptable technologies.

This volume is a record of progress using the matrix approach to investigate the environmental and health effects of one component of one synfuels technology.

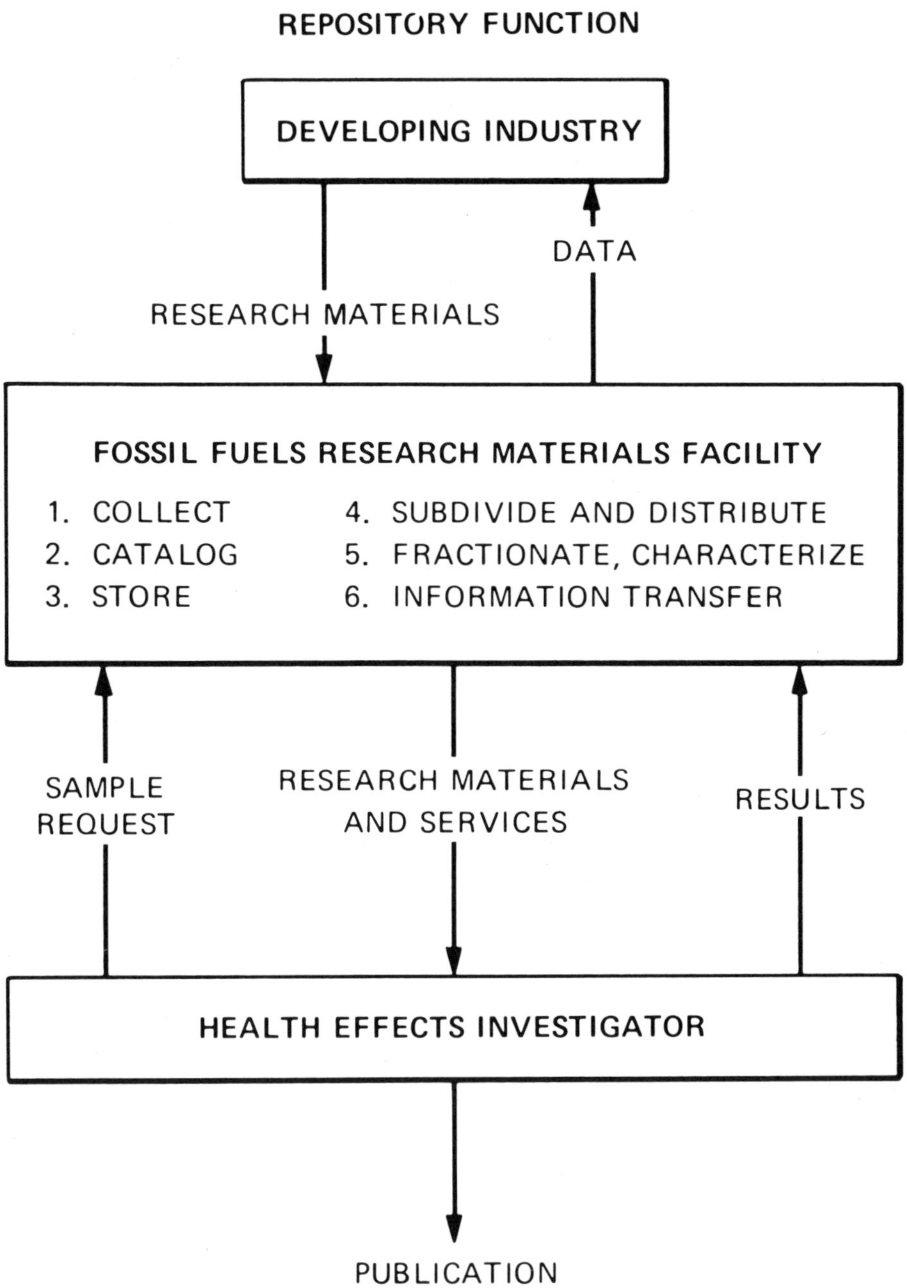

Figure 1. Functions of the Research Materials Facility

REFERENCES

1. M. R. Guerin, "The Integrated Approach to Chemical-Biological Analysis," Second Symposium on Process Measurements for Environmental Assessment, Atlanta, GA, Feb. 25-27, 1980. Proceedings in press.

2. W. H. Griest, D. L. Coffin, and M. R. Guerin, "Fossil Fuels Research Matrix Program," ORNL/TM-7346, Oak Ridge National Laboratory, Oak Ridge, TN, June, 1980.

3. D. L. Coffin, M. R. Guerin, and W. H. Griest, "The Interagency Program in Health Effects of Synthetic Fossil Fuels Technologies: Operation of a Materials Repository," Proceedings of the Symposium on Potential Health and Environmental Effects of Synthetic Fossil Fuel Technologies, Gatlinburg, TN, 1978, CONF-780903, 153-156 (1979).

CHAPTER 2

PARAHO OPERATIONS

R. N. Heistand, Development Engineering Inc., Paraho Development Corporation, Grand Junction, Colorado 81501

INTRODUCTION

Brief History

Paraho oil shale retorting operations began with the Paraho Oil Shale Demonstration. That demonstration, a $10-million, 3-year project having 17 industrial participants, has been described previously.[1,2] The main purpose of that project, the demonstration of the reliability of Paraho retorting technology in processing oil shale, has been achieved. Another achievement of that project was the refining of 10,000 barrels of crude shale oil into military fuels.

Based in part upon the achievement of that refining run, the Office of Naval Research (ONR), Department of Navy, and the Department of Energy (DOE) contracted with Paraho to continue research and development of surface retorting technology during 1977 and 1978.[3]

The crude shale oil produced during those operations has been shipped for refining into military fuels at the Standard Oil Company of Ohio's Toledo refinery.

The Paraho accomplishments attained during the ONR-DOE two-year operations can be summarized as follows:

- Production of more than 100,000 barrels of crude shale oil.
- Continouous period of operation exceeding 100 days.
- Maintained a service factor exceeding 90% for a two-year period.
- Ability to retort rich shale, + 35 gallons per ton.
- Ability to produce low sulfur gas ($<$ 500 ppm H_2S).
- Ability to operate until scheduled turn-around.

The Paraho Process

The Paraho process has been operated using both pilot and semi-works size retorts in various modes of operation. The process, retorts, and operating modes have been described previously.[4,5]

In the Direct Mode, combustion required to produce the heat for retorting occurs within the retort (see Figure 1). Raw shale enters the top of the retort and is preheated in the mist formation zone by the gases carrying the oil mist out of the retort. The preheated shale next enters the retorting zone where the hot gases rising from the internal combustion heats the shale to retorting temperature. Upon retorting, the solid organic kerogen in the shale breaks down to gas, oil, and coke. The gas and oil are swept upward with the hot gases and the coke remains on the retorted shale. As the retorted shale enters the combustion zone, some of this coke is burned with the air introduced through the gas/air distributor. Only enough air is introduced to produce enough heat for the retorting process. Recycle gas is added with the air to insure even distribution across the bed and to control the flame temperature. Below the middle distributor, the retorted shale enters the cooling zone where it is cooled by recycle gas rising from the bottom distributor. In the Paraho retort, flows are counter current with gases rising and shale moving downward under the force of gravity and controlled by the grate. These flows produce the hot combustion and retorting zones in the middle with heat exchange both top and bottom so that both the gases and retorted shale are relatively cool when they leave the retort.

RECENT OPERATIONS

Paraho Development Corporation has been involved in numerous oil shale operations during the past year. These involve the following special projects involving use of Anvil Points facilities: laboratory studies, pilot plant retorting operations; consulting and special data compilation. many of the activities will be discussed briefly in this section.

Oil storage

The 10,000-barrel tanks, constructed to store Paraho oil during the ONR operations, have been cleaned and prepared to store oil under contract with the DOE. Presently, most of those tanks are in use storing oil produced by several in-situ processes.

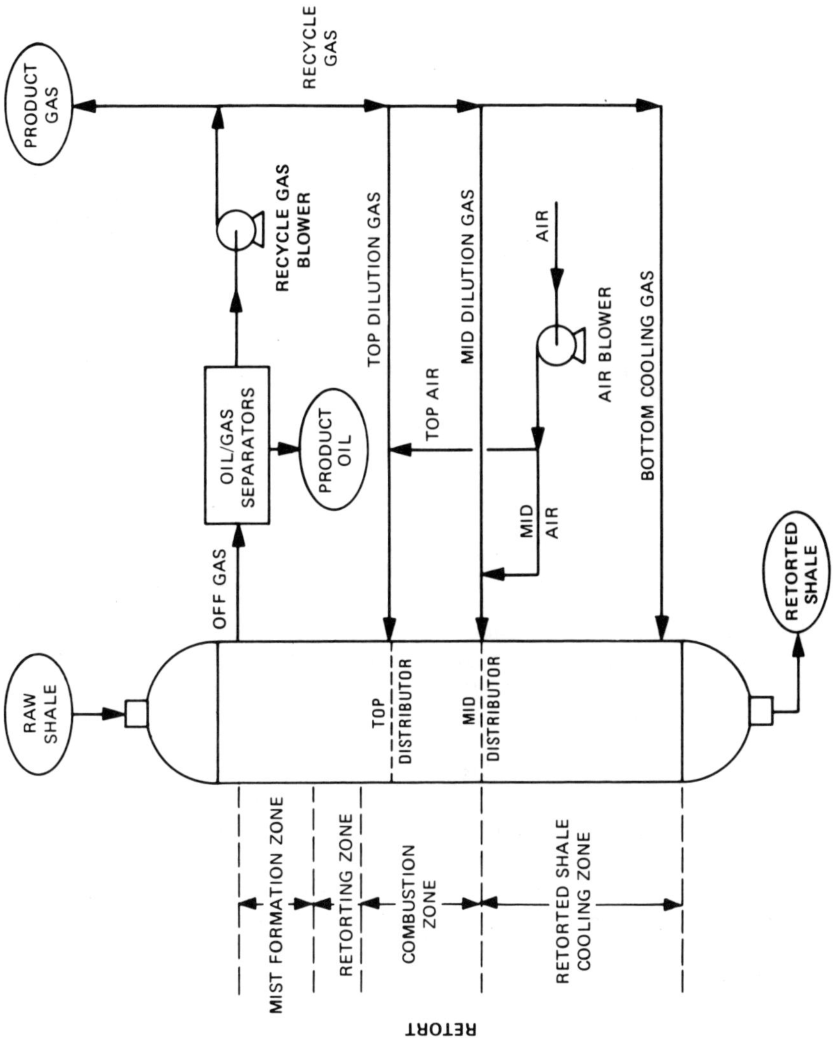
RAW
SHALE
PRODUCT
GAS
OFF GAS
OIL/GAS
SEPARATORS
RECYCLE GAS
BLOWER
RECYCLE
GAS
PRODUCT
OIL
MIST FORMATION ZONE
RETORTING ZONE
COMBUSTION
ZONE
RETORTED SHALE
COOLING ZONE
RETORT
TOP
DISTRIBUTOR
MID
DISTRIBUTOR
TOP DILUTION GAS
TOP AIR
MID DILUTION GAS
MID
AIR
AIR
AIR BLOWER
BOTTOM COOLING GAS
RETORTED
SHALE

Laboratory Studies

In addition to performing some specialized studies for various clients, the Paraho Laboratory is currently engaged in a core examination project. This entails receipt of core, splitting by sawing, crushing and sample preparation, Fischer Assay examination, and data processing.

Pilot Plant Retorting Operations

During this past year, the two-and-one-half-foot inside diameter Paraho pilot plant retort was used to process foreign shales under contract. Shale from Israel has been processed. A contract has been signed to process shale from Morrocco. These pilot retorting operations have been successful and it has been reported that both nations are considering scale-up towards the construction and operations of pioneer commercial plants. Unfortunately, in this instance, the United States of America has lost its pioneering position. Apparently we're too concerned with possible adverse environmental impacts, and too unconcerned with the real energy shortage, high costs of energy, along with the resulting chaos to our national economy. The U.S. needs well-planned, stable environmental and energy policies to regain its pioneer position.

Retorted Shale Research

Retorted shale research studies performed by Paraho at Anvil Points were summarized in a paper presented at the 13th Oil Shale Symposium, Colorado School of Mines, Golden, CO.[6] These research studies included extensive laboratory testing, field compaction studies using 15,000 tonnes of retorted shale, and field infiltration studies consisting of two ponds about 30 m diameter. The results of this research (see Table I) showed that: Paraho retorted shale can be compacted to dry densities ranging from 1550 to 1600 kg/m^3; the retorted shale, properly treated, has the properties of a low-grade cement (compressive strength equals 1380 kPa with nothing other than water added); the retorted shale can be made impervious to water flow (rate $\leq 1 \times 10^{-7}$ cm/sec); leaching autoignition and dusting do not pose any problems.

Data Compilation

Paraho, a leading oil shale technology and development company, has also been a leader in the cooperation with government agencies, universities, and research laboratories in obtaining samples and information in environmental and health related areas. The samples stored in the ORNL Repository are evidence of this cooperative effort. The sources of the ORNL

Table I. Summary of Retorted Shale Research

Compaction, dry density	1400-1600 kg/m^3 (87 to 106 pcf)
Permeability permeation rate	1×10^{17} cm/sec (0.1 ft/yr)
Strength unconfined compression	1480 kPa (215 psi)
Classification	
type	silty - gravel
size	30 - 50% > 4.76 mm (4 mesh) 25 - 35% < 0.074 mm (200 mesh)
Leaching, Auto-ignition and Dusting	No problem in laboratory or field.

Repository samples and their relationships within the process are shown in Figure 2.

As a result of Paraho's cooperative effort, more than 30 different groups have obtained samples and have published or presented over 100 papers containing Paraho data. In 1979 Paraho was awarded a contract to compile this data into eight categories (see Table II). This compilation consists of collection and collating papers, preparing a bibliography, collating and compiling data, reviewing the compiled data and issuing reports.

Table II. Oil Shale Data Evaluation Panels

1. Process Characterization
2. Air Quality
3. Water Quality
4. Land Disturbance
5. Ecology
6. Health Effects
7. Occupational Health
8. End Use

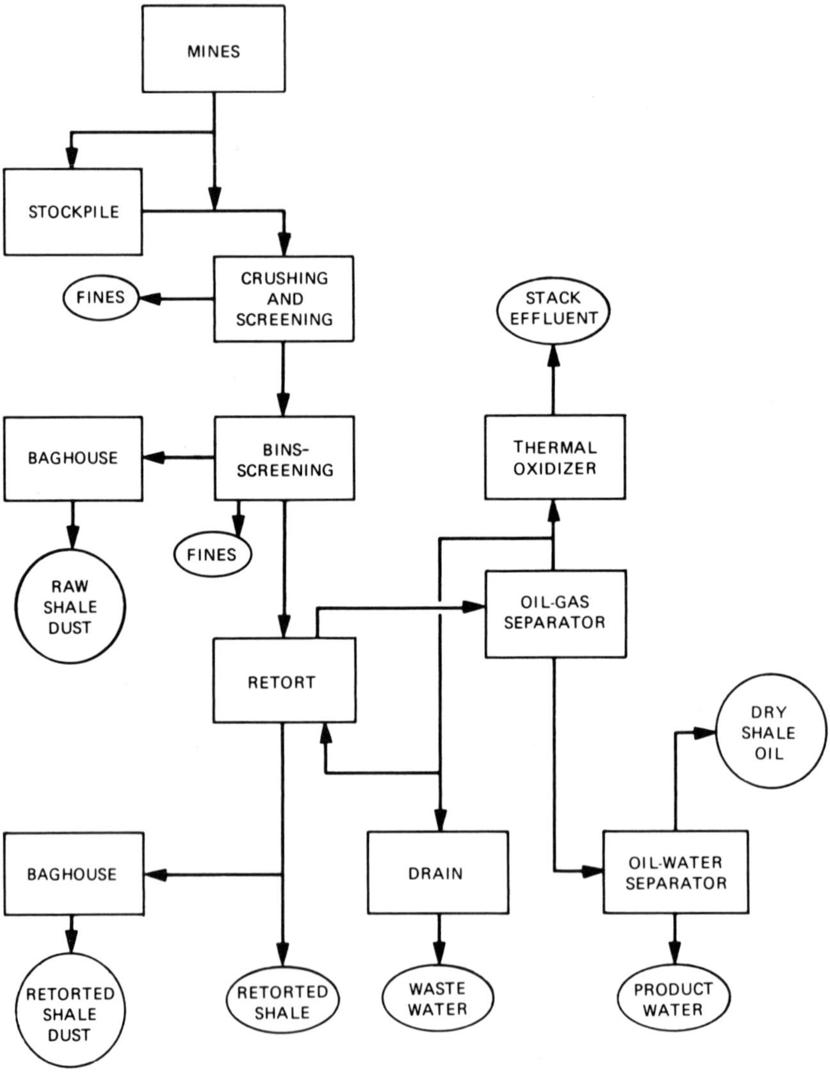
MINES
STOCKPILE
CRUSHING AND SCREENING
FINES
STACK EFFLUENT
BAGHOUSE
BINS-SCREENING
THERMAL OXIDIZER
FINES
RAW SHALE DUST
OIL-GAS SEPARATOR
RETORT
DRY SHALE OIL
BAGHOUSE
DRAIN
OIL-WATER SEPARATOR
RETORTED SHALE DUST
RETORTED SHALE
WASTE WATER
PRODUCT WATER

Figure 2

CONCLUSIONS

1. Paraho is pleased to assist the health effects researchers in obtaining valid samples for these important studies. The work of this symposium, the coordination of research findings, the open discussion of experimental results and procedures, and the compilation of the valid results - should provide a broad and valid data base which should serve as a model for future reference.

2. Paraho Development Corporation continues in its effort to promote the orderly and systematic development of a commercial oil shale industry. To this goal, it has continued to work with governments (both U.S. and foreign), private industry, and research groups.

REFERENCES

1. J. B. Jones, Jr., "The Paraho Retort," 9th Oil Shale Symposium, Colorado School of Mines, Golden, CO, 1976.

2. J. B. Jones, Jr., "Technical Evaluation of the Paraho Process," 11th International Conference on Mechanical Engineering, Techion, Haifa, Israel, 1977.

3. J. B. Jones, Jr., and R. N. Heistand, "Recent Paraho Operations," 12th Oil Shale Symposium, Colorado School of Mines, Golden, CO, 1979.

4. H. Pforzheimer, H., Jr., "Paraho - New Prospects for Oil Shale," Chemical Engineering Progress, 70(9), 62-65 (1974).

5. R. N. Heistand and J. B. Jones, Jr., "Recent Paraho Operations," Symposium on Health Effects Investigation of Oil Shale Development," U.S. Environmental Protection Agency, Research Triangle Park, NC, 1979.

6. R. N. Heistand and W. G. Holtz, "Retorted Shale Research," 13th Oil Shale Symposium, Colorado School of Mines, Golden, CO, 1980.

ACKNOWLEDGEMENT

This work was performed at the Department of Energy Anvil Points Oil Shale Research Facility situated on the Naval Oil Shale Reserves Nos. 1 and 3 located near Rifle, Colorado.

CHAPTER 3

RESULTS OF THE SOHIO REFINING RUN

D. L. Cawein, Director, Engineering, Health, Environmental Affairs, Standard Oil Company, Cleveland, Ohio 44115

INTRODUCTION

In 1976, negotiations were finalized between private industry and the U. S. government for large scale production and testing of transportation fuels derived from oil shale that would meet military specifications. The Paraho Development Corp. contracted with the United States Navy Office of Naval Research and DOE to produce up to 100,000 barrels of shale oil at the Anvil Points, Colorado research facility. Retorting operations commenced in January, 1977, and continued through September, 1978. Included in these operations was an extensive program for data acquisition including environmental monitoring.

Concurrently Paraho contracted with Sohio to conduct a two-stage program to refine the Paraho-produced raw shale oil into specification military fuels. The first stage consisted of pilot plant testing using a configuration adoptable to existing and available facilities at Sohio's Toledo, Ohio refinery. The second stage consisted of the full scale refining of the raw shale oil in commercial equipment and the shipping of products to various government and industry fuel testing locations.

Sohio confirmed in its pilot testing program that its commercial facilities with minor modifications could produce specification products. 73,000 barrels of shale oil were processed during the winter of 1978-79 as discussed in more detail below.

PILOT TESTING

The unique nature of raw shale oil requires special considerations in handling and processing. Table 1 summarizes some typical inspection of raw shale oil and hydrotreated shale oil. Important considerations include:

Table I. Shale Oil Quality

	Raw Shale Oil	Hydrotreated Shale Oil
Gravity, °API	20.4	33.8
Dist., % @ 400°F	5	13
% @ 650°F	28	52
% @ 900°F	62	90
Pour Point, °F	85	85
Visc., SSU @ 100°F	213	46
Sulfur, Wt.%	.70	.002
Nitrogen, Wt.%	2.02	.33
Oxygen, Wt.%	1.38	.29
Metals, PPM		
- Arsenic	7.5	.2
- Nickel	2.4	.05
- Iron	53	.04
- Vanadium	.2	.006

(a) High nitrogen - This is the major area of concern with shale oil as compared with most crude oils. The 2 wt.% of nitrogen is distributed rather evenly throughout the total boiling range of shale oil and represents an increase of 10 to 20 times as much as is found in crude oil fractions. The nitrogen compounds create a source of major instability to fuel products and substantially increase NO_x in exhaust flue gases. It also creates a major poisoning of many catalysts used in various refinery processes. Proven technology is available to remove nitrogen by hydrotreating.

(b) Sulfur - The sulfur level is low enough to cause raw shale oil to be considered equivalent to low-sulfur crude oil. With hydrotreating necessary to remove nitrogen, essentially all of the sulfur will be converted to H_2S.

(c) Metals - The metals content of shale oil is substantially higher than most crude oils. These organometallic compounds are adsorbed on the catalyst used in hydrotreating raw shale oil and do complicate the hydrotreating process due to catalyst deactivation and plugging of the catalyst bed.

(d) <u>Boiling Range</u> - Raw shale oil is substantially heavier than most crude oils and contains a much lower yield of gasoline, jet fuel, and diesel fuel. The yield of these light products is almost doubled by hydrotreating and the remaining 650° F+ residue after hydrotreating is an excellent cracking feedstock.

(e) <u>Pour Point</u> - Both raw shale oil and hydrotreated shale oil have very high pour points and are much more viscous than most crude oils. This complicates the pumping and handling of shale oil which, however, can be accommodated at some additional cost.

The net result of the pilot testing showed the following major conclusions:

(a) The severity of hydrotreating as set by the operating limits of the commercial equipment at Sohio's Toledo Refinery would not accomplish sufficient removal of nitrogen to meet military fuel specs. However, when followed by acid-treating, all military specs could be met. Acid-treating would probably not be required in a hydrotreating facility specifically designed for raw shale oil.

(b) The yield of gasoline, jet fuel, and diesel fuel would be approximately 70% of the hydrotreated product.

(c) Additional facilities would be required in the refinery run to guard against plugging of the hydrotreating catalyst due to suspended solids and organic-metallic compounds.

REFINERY RUN

One of the processing units at Sohio's Toledo Refinery is a 10,000 BPD (barrel per day) two-stage Hydrocracker. It has been in operation since 1962 and is primarily used to upgrade distillate fuels to gasoline. Sohio was willing to rent the use of this unit for a 30-day period. Based on the pilot testing results, calculations indicated that the first stage plus the fractionation section could, with minor modifications, be capable of hydrotreating about 3000 BPD of raw shale oil. Other modifications at the Toledo Refinery included tank car unloading and loading facilities, tankage for the raw shale oil and hydrotreated products, and the addition of a new acid treater and clay contactor (using refinery surplus equipment).

The general processing plan is shown in Figure 1. This consists of:

TOLEDO REFINERY BLOCK FLOW

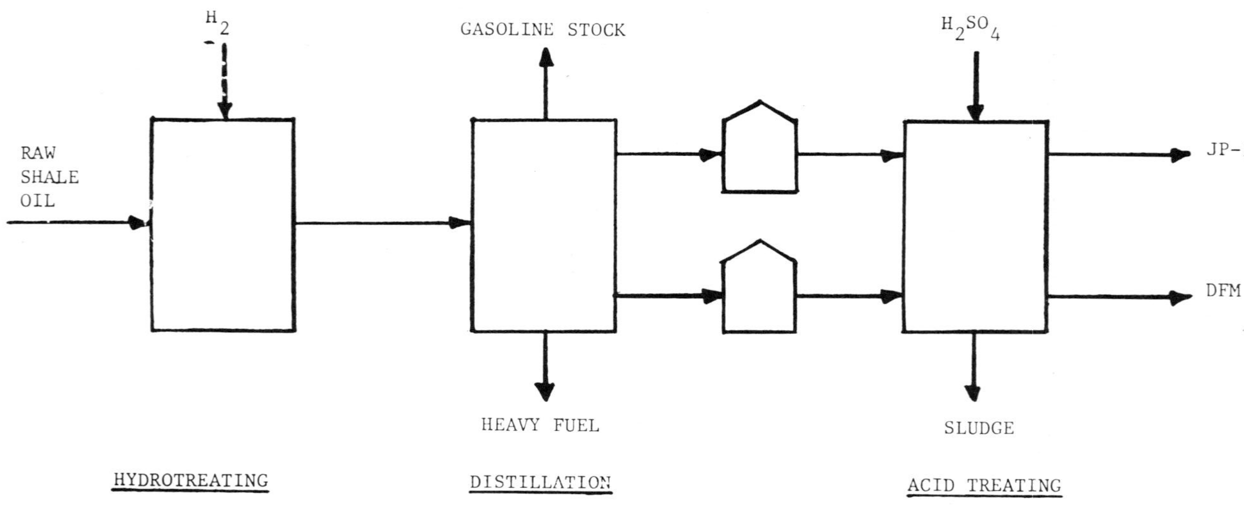

Figure 1

Hydrotreating - A catalytic reaction using commercially available metallic oxide catalyst in a hydrogen atmosphere at elevated temperatures and pressure to convert nitrogen, sulfur, and oxygen compounds into NH_3, H_2S and H_2O.

Distillation - Separation by fractional distillation into gasoline, jet fuel, diesel fuel and residue.

Acid Treating - The removal of any unconverted nitrogen compounds by contacting with sulfuric acid, followed by percolation through clay.

During the summer and fall of 1978 the 88,000 barrels of raw shale oil that had been produced by Paraho at Anvil Points, Colorado were transported to Toledo Refinery. The various modifications were completed at Toledo Refinery to accommodate the refining run, including a several week shutdown of the Hydrocracker to convert to a single stage operation and install a new charge of catalyst. On November 4, 1978, the hydrotreating run started and continued through December 4, 1978, by which time 73,000 barrels of raw shale oil had been processed. The run was terminated at this point due to an excessive pressure drop in the guard bed which could not be remedied within the contractual time limitations.

The detailed processing flow during the refinery run is shown in Figure 2. Raw shale oil was pumped from storage through a 20 micron mechanical filter (to remove suspended solids) and into the Hydrocracker unit. The oil was preheated and passed through the guard bed for primary removal of iron and arsenic. THe guard bed was filled with alumina extrudate. The oil was then mixed with a hydrogen rich gas stream and brought up to reaction temperature in the furnace. This liquid-gas mixture at a pressure in excess of 100 atmospheres flowed down through several beds of catalyst in the reactor. Heat of reaction was dissipated by adding cool hydrogen-rich gas between the beds. Reactor effluent was cooled and mixed with steam condensate for partial removal of ammonia and hydrogen sulfide. After separation of the gas and foul condensate, the oil was stripped of any remaining ammonia, hydrogen sulfide, water and light hydrocarbons in a low pressure stripper tower. The stripped hydrotreated product was then separated in a 50 tray, reboiled and refluxed fractionating tower into the four basic products: unstabilized gasoline stock, jet fuel, diesel fuel, and a residual product. The gasoline stock was sent elsewhere in the refinery for stabilization and the residual loaded directly into tank cars or sent elsewhere in the refinery as a cracking feedstock.

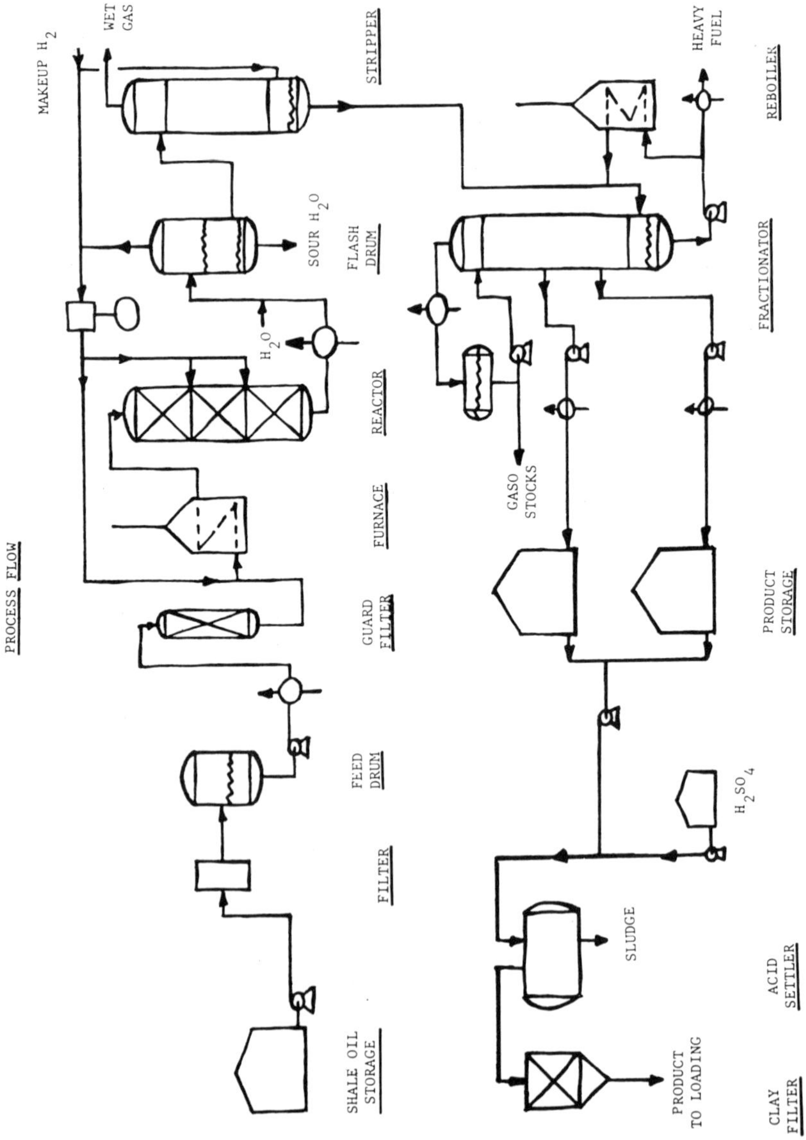
PROCESS FLOW
SHALE OIL STORAGE
FILTER
FEED DRUM
GUARD FILTER
FURNACE
REACTOR
FLASH DRUM
STRIPPER
MAKEUP H_2
WET GAS
SOUR H_2O
H_2O
GASO STOCKS
FRACTIONATOR
REBOILER
HEAVY FUEL
PRODUCT STORAGE
H_2SO_4
SLUDGE
ACID SETTLER
CLAY FILTER
PRODUCT TO LOADING

Figure 2

Both the jet fuel and diesel fuel sidestreams were stripped of light boiling fractions to meet flash point specifications and then run down to intermediate storage. They were then pumped one at a time to the acid treating section from which they were directly run down to tank cars. The treating consisted of intimate contacting with 93% H_2SO_4 using a static mixer. Nitrogen compounds react quickly to form a viscous acid sludge. Most of the sludge were removed in a coalescer followed by percolation through an Attapulgus clay filter.

Provisions were made to produce a small proportion of jet fuel meeting JP-8 specifications by minor changes to the fractionating conditions. During the period the jet fuel was run down directly from the tower through the acid treating section and into the shipping car. Provisions had been made to produce a small amount of JP-4 in a similar fashion by blending a portion of the gasoline stock with the jet fuel sidestream. However, JP-4 was not produced because a steady flow of gasoline stock necessary for blending purposes could not be attained.

REFINERY RUN RESULTS

A summary of product yields is shown in Table II. The as produced yield of both jet fuel and diesel fuel was substantially lower during the refinery run than during the pilot run. However, this was due to a poorer distillation efficiency with the refinery fractionator than with the pilot facilities. An examination of the distillation data for the refinery products indicates that a substantial amount of jet fuel was contained in the front end of the diesel fuel and the back end of the diesel fuel was left in the residual product. This is further substantiated by comparing the distillation curve of the hydrotreated product before fractionation of both the pilot plant and the refinery run. As shown on Figure 3, both distillation curves are essentially the same. Thus with fractionation equipment specifically designed for high efficiency separation, yields comparable to the pilot results would be possible. Of more importance for commercial operation is the fact that the measured hydrogen consumption during the refinery run was substantially lower than during the pilot run. The refinery hydrogen consumption is believed more accurate because of measurement and leakage problems in the pilot equipment.

The quality of jet fuel (JP-5) produced during the refinery run was very comparable to that produced during the pilot run as shown in Table III. The nitrogen was slightly lower but still much too high to meet the thermal stability requirements. However, acid treating removed essentially

HYDROTREATED SHALE OIL

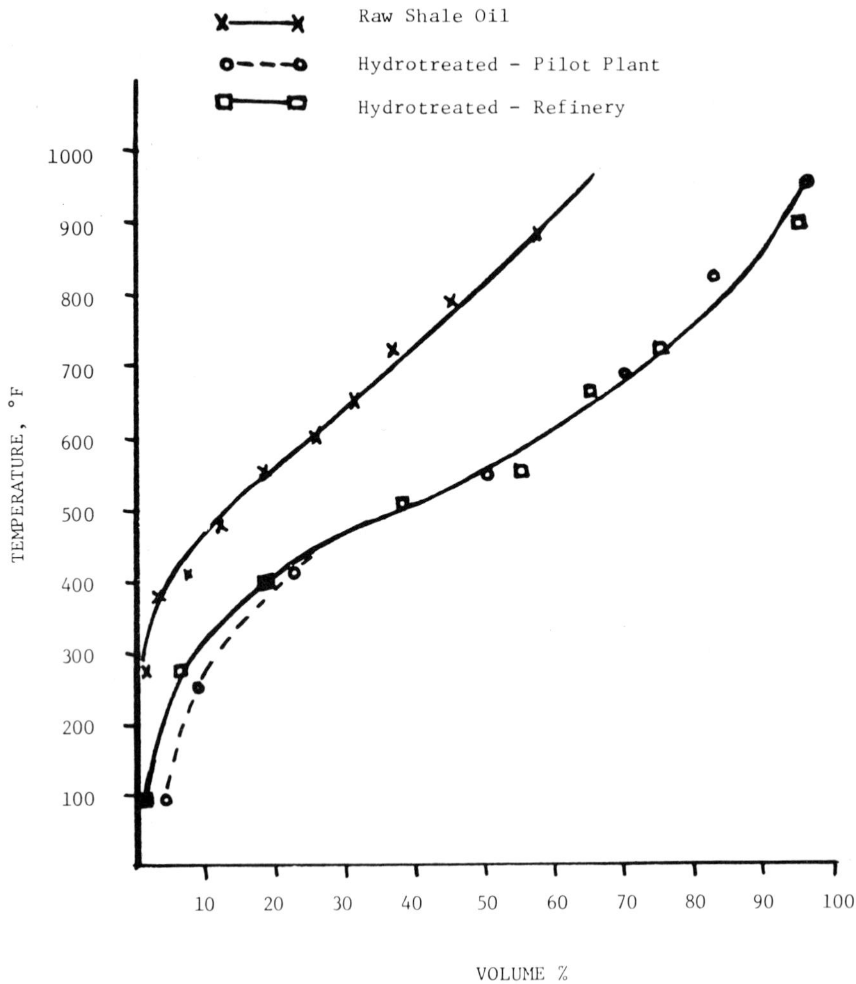

Figure 3

Table II. Product Yields

	Pilot Plant	Refinery
Gasoline, Vol. %	11.0	12.0
JP-5, Vol. %	25.3	13.7
DFM, Vol. %	34.5	25.9
Heavy Fuel, Vol. %	35.2	50.9
Liquid Products, Vol. %	106.0	102.5
H2 Consumed, SCF/BBL.	1500	1050

Table III. Jet Fuel (JP-5) Quality

	Pilot Plant	Refinery
	Untreated	
Gravity, °API	41.8	42.7
Dist., IBP, °F	356	370
50%	415	400
EP	477	480
Flash Pt., °F	157	158
Freeze, °F	-52	-57
Paraffins, Vol. %	43.9	42.5
Naphthenes, Vol. %	33.1	36.1
Aromatics, Vol. %	23.0	21.5
Nitrogen, PPM	3200	2900
ASTM Gum	12.2	4+
JFTOT @ 500°F	Fail	Fail
	Acid Treated	
Aromatics, Vol. %	20.9	21.8
Nitrogen, PPM	8	0.6
ASTM Gum	0.6	1.4
JFTOT @ 500°F	Pass	Pass

all of the remaining nitrogen compounds resulting in a very stable product.

The quality of diesel fuel produced during the refinery run was considerably lighter than that produced during the pilot run as shown in Table IV. However, both met distillation specifications. The nitrogen content during the refinery run was lower but still did not meet thermal stability requirements. After acid treating, stability was very good. Combustion quality of the diesel fuel should be excellent as indicated by the high cetane index.

Table IV. Diesel Fuel Marine (DFM) Quality

	Pilot Plant	Refinery
	Untreated	
Gravity, °API	33.4	36.8
Dist., IBP, °F	507	396
50%	553	512
EP	628	582
Flash Pt., °F	290	162
Pour, °F	15	0
Cetane Index	50	52
Nitrogen, PPM	4000	3300
Stability	-	8.3
	Acid Treated	
Nitrogen, PPM	90	4
Aromatics, Vol. %	32	29
Cetane Index	55	54
Stability	0.4	0.5

The quality of the residual product as shown in Table V was substantially better than any specifications for clean, low-sulfur, high pour #6 fuel oil. In commercial operation, the residual product with the low metals content would be much more valuable as a cracking feedstock than as a boiler fuel.

CONCLUSIONS

(a) Proven processing technology is available to refine raw shale oil into military quality transportation fuels.

Table V. Heavy Fuel Quality

	Pilot Plant	Refinery
	Untreated	
Gravity, °API	30.3	29.6
Dist., IBP, °F	689	331
50%	830	732
EP	1000	1032
Pour, °F	105	80
Vis @ 210°F, CST	6.45	2.00
Asphaltenes, Wt. %	0.24	0.06
Aromatics, Vol. %	43	37
Sulfur, PPM	20	20
Nitrogen, PPM	3300	4400
Metals, PPM		
- Arsenic	0.1	0.4
- Iron	0.9	0.1
- Vanadium	0.4	0.02

(b) These finished products will be compatible with and very similar in physical and chemical characteristics to products refined from crude oils.

(c) There is a very close correlation between pilot plant and refinery yields and quality.

(d) Problem areas can be identified and solutions pursued at the pilot plant level.

(e) Supplementary pilot studies confirm that satisfactory product quality can be met by high severity hydrotreating in facilities specifically designed for raw shale oil, eliminating the need for the acid treating step.

CHAPTER 4

SAMPLE MANAGEMENT AND CHEMICAL CHARACTERIZATION OF THE PARAHO/SOHIO/U.S. NAVY CRUDE AND REFINED SHALE OIL SUITE*

W. H. Griest, M. R. Guerin, L. B. Yeatts, Jr., and B. R. Clark, Oak Ridge National Laboratory, Oak Ridge, Tennessee 37830

INTRODUCTION

The role of the EPA/DOE Fossil Fuels Research Materials Facility (the "Chemical Repository") in this study is to actively support and coordinate the health effects investigations of this oil shale development by acting as a central point for acquisition, cataloging, storage, and distribution of referenceable materials for study. In addition, selected materials are chemically fractionated for bioassay and are characterized in additional detail beyond that of routine stability monitoring. Finally, the Facility is acting as an information transfer agent to insure that the results of the investigations are made available to the oil shale industry and the funding agencies.

This chapter is concerned with the sample management and characterization aspect of the Facility's role in this study.

SAMPLE ACQUISITION AND INVENTORY

The first concern of the Facility in this investigation was to obtain shale- and petroleum-derived materials relevant for study. Through the cooperation of the U.S. EPA, U.S. DOE, U.S. Navy, and The Standard Oil Company of Ohio (SOHIO), arrangements were made to have samples of the crude shale oil, process materials, and final, finished jet and diesel fuel products collected and sent to the Facility. Lt. Cdr. L. J. Jenkins of the U.S. Navy Medical Research Institute,

*Research sponsored jointly by the U. S. Environmental Protection Agency and the U. S. Department of Energy under contract W-7405-eng-26 with the Union Carbide Corporation.

Table I. Inventory of Crude and Refined Shale Oils and Petroleum Equivalents

Repository Sample Number	Date	Description	SOHIO Reference Number	Quantity (Gal)
4601	11-03-78[a]	Crude Shale Oil	CSO-556	53
4602	11-29-78[a]	Hydrotreated Shale Oil	C5HTSO-554	53
4603	11-21-78[a]	Weathered Gasoline Feedstock	WGFS-55	4.5
4604	11-22-78[a]	JP-5 Precursor	PREJP5-555	53
4605	11-27-78[a]	JP-8 Precursor	PREJP8-555	53
4606	11-21-78[a]	DFM Precursor	PREDFM-555	53
4607	11-30-78[a]	Hydrotreated Residue	HTR-555	53
4608	1-24-79[a]	JP-5 Product	FINJP5-554	53
4609	11-27-78[a]	JP-8 Product	FINJP8-554	53
4610	2-21-79[a]	DFM Product	FINDFM	53
4612	-[b]	Acid Sludge from DFM Treatment	-	4.5
4613	1-19-79[c]	Petroleum JP-4	-	55
4614	1-10-79[c]	Petroleum JP-5	-	53
4615	1-10-79[c]	Petroleum JP-8	-	53
4616	1-10-79[c]	Petroleum DFM	-	3

[a]Date of sampling.
[b]Date not supplied.
[c]Date of receipt.

Wright Patterson Air Force Base, Ohio, provided the petroleum-derived jet and diesel fuels which the U.S. Navy is using as a "conventional" reference point in their toxicity studies.

A major objective was to obtain the materials in an uncontaminated state. For this reason, type 304 stainless steel 55 gal drums were obtained by the Facility, were cleaned, and were shipped to the SOHIO refinery at Toledo, Ohio for direct collection of the shale-derived samples and return shipment. Smaller (5 gal) samples of highly volatile or corrosive process materials were collected in Teflon-lined steel pails, also provided by the Facility. In the case of the petroleum derived fuels, shipment was made to the Facility in carbon steel drums lined with an epoxy resin which the Navy has found effective for fuel storage. Both sets of samples were received at the Facility in the winter of 1978-1979. They were sampled in the spring of 1979 (see below) and are now being stored at 4°C in the storage Repository of the Facility.

The current inventory of the bulk volumes of these materials is listed in Table I. The information shown includes the unique identifying Repository Sample Number assigned to each material, the descriptive name of the material, the reference number used by the SOHIO refinery, and the current bulk sample amount stored at the Facility. The suite of materials includes the crude shale oil (No. 4601), six process materials (Nos. 4602-4607), three finished jet fuel and diesel fuel marine products (Nos. 4608-4610), and a byproduct (No. 4612). Included in the suite are four petroleum-derived jet fuels and diesel fuel marine (Nos. 4613-4616) which the U.S. Navy is using as a "conventional fuel" reference point for comparison with the shale-derived fuels. The Petroleum JP-4 (No. 4613) is not currently being studied because of the decision not to produce a shale JP-4, but it is available for future use.

SAMPLE ALIQUOTTING

The sample aliquotting protocol was designed to maximize sample homogenicity, aliquot uniformity, and sample stability, and to minimize the possibility of sample contamination and degradation. The main features of this protocol were as follows:

a) On the day of sampling of a particular material, the drum lid was replaced with a special stainless steel lid fitted with a stainless steel ball valve and an adjustable air leak to allow direct withdrawal of the sample from the drum to minimize contamination in handling.

b) The sample and aliquot-to-aliquot uniformity was maximized by solar heating of the drums (accomplished by spraying a 180° sector of the drum exterior with flat-black paint and orienting this painted side of the drum toward the sun) to raise the material to a temperature above its pour point. Thorough mixing was accomplished by rotation of the drum after inserting two six inch lengths of type 304 stainless steel pipe into the drum to facilitate agitation. The lighter, low viscosity materials such as the jet and diesel fuels were sampled in the mornings, while the heavier, more viscous crude and hydrotreated shale oil and hydrotreated residue were sampled in the afternoons to allow longer solar heating periods. These measures were effective for homogenizing most of the samples. However, the drums of Crude Shale Oil, Hydrotreated (HDT) Shale Oil, and HDT Residue all contained a layer of a thick gel material which was not completely dissolved by this treatment. Although the material was not completely homogenized, the aliquots appeared to be uniform from sample to sample.

c) To minimize contamination, the samples were withdrawn directly into the sampling containers which had been rinsed with methanol, dried, and pre-rinsed with a portion of the sample material itself. After sampling, the headspace of each container was briefly flushed with argon to exclude oxygen and to provide an inert blanket.

d) To minimize possible photodegradation and sample-container interaction, the sampling containers were foil-covered or amber-colored borosilicate glass bottles of 0.125, 0.5, and 1 l volumes. Borosilicate glass is the preferred[1] container material. Soft glass has been shown[2] to interact with hydrocarbon fuels. For larger sample requirements, uncoated carbon steel drums (5 gal capacity) also were used. These drums were specially ordered to omit the usual rust-inhibitor inner surface coating which could contaminate the sample.

e) The samples were shipped immediately by air express to minimize the delay between aliquotting by the Facility and receipt by the investigators participating in the study.

f) Sample aliquots not immediately distributed were stored in the dark at -30°C to minimize degradation.

SAMPLE DISTRIBUTION FOR MATRIX-APPROACH STUDY

Although the most of the samples were distributed soon after the aliquotting was conducted, additional aliquots have been distributed since then to other investigators. The current distribution of this suite of materials is shown by the study matrix in Table II, where the materials and studies form adjacent axes, and the investigators (Key to investigators in Table III) comprise the matrix.

The matrix indicates that a wide variety of investigations and talent from institutions across the country is being focused on a common set of materials. Laboratories of the U. S. government, private industry, and private and public universities are participating. Studies of the materials range from physical and chemical characterization to biological and ecological effects, including toxicity, mutagenesis, and carcinogenesis. Additional investigators will be added as the study continues.

SAMPLE STABILITY MONITORING

A continuing concern of the Facility is the stability of the materials under storage. In order for data obtained with materials shipped months or years apart to be considered comparable, the stability of the materials must be monitored and characterized to demonstrate minimal change.

A battery of routine stability tests has been set up to characterize several parameters of stability, based on reports[2-6] of changes in the physical and chemical properties of crude and refined synfuels in accelerated aging tests. The battery of tests includes the following:

a) Direct gas chromatographic (GC) profile of the material to detect changes in the major chromatographable organic constituents which would occur with evaporation, degradation, or contamination of the major volatile organic constituents.

b) Infrared spectrum (IR) of the material to detect changes in the organic functionality, particularly in the $\sim$ 1700 cm^{-1} spectral region for absorption by carbonyl groups. Increases in carbonyl absorption band intensities correlate with aging and gum formation.[4,6] Contamination also may be detected if the contaminant has a unique, well-defined spectrum different from that of the sample.

c) Viscosity and density measurements to detect aging. The viscosity increases[3,5,6] upon aging.

Table II. Study Matrix of the Paraho/SOHIO Crude and Refined Shale Oil Suite[a]

Repository Number	Research Material	Chemistry: Comprehensive Analysis	PAH Analysis	Stability Screening	Special Studies	Chemical Class Fractionation	Biology, Ecology: Mouse Skin Carcinogenesis	Bacterial Mutagenesis	Drosophilia Mutagenesis	Acute Oral Toxicity to Mice	Pond Ecosystem Effects	Marine Ecosystem Effects	Combustion Products Mutagenicity	Cilia Toxicity	Mouse Lung Tumorigenicity, Skin & Eye Irritation, Toxicity	Mice Interperitoneal Tumorigenesis	Trout Embryo Larvae Hatching & Morphology
4601	Crude Shale Oil	1	2,5	15	15	16	3–5	5–7	8,9	10	11	—	—	13	17	18	19
4602	HDT Shale Oil	1	2,5	15	15	16	3–5	5–7	8	10	11	—	—	13	17	18	—
4603	Weathered Gas Feedstock	1	—	15	—	16	—	6	8	—	—	—	—	—	—	—	—
4604	JP-5 Precursor	1	—	15	—	16	—	6	8	—	—	—	—	—	—	—	—
4605	JP-8 Precursor	1	—	15	—	16	—	6	8	—	—	—	—	—	—	—	—
4606	DFM Precursor	1	2	15	—	16	3,4	6	8	—	—	—	12	—	—	—	—
4607	HDT Residue	1	2,5	15	15	16	3–5	5,6	8	10	11	—	—	13	17	—	—
4608	JP-5 Product	1	5	15	—	16	3–5	5,6	8,9	—	11	—	—	—	17	—	—
4609	JP-8 Product	1	5	15	—	16	3–5	5,6	8	—	—	—	—	—	17	—	—
4610	DFM Product	1	2,5	15	15	16	3–5	5,6	8	—	11	14	12	13	17	18	—
4612	Acid Sludge	1	—	—	—	16	—	6	8	—	—	—	—	—	—	—	—
4614	Petroleum JP-5	—	5	15	—	16	4,5	5,6	8,9	—	11	—	—	—	17	—	—
4615	Petroleum JP-8	—	5	15	—	16	4,5	5,6	8	—	—	—	—	—	17	—	—
4616	Petroleum DFM	—	5	15	—	16	3	5,6	8	—	11	14	—	13	17	18	—

[a]Key to investigator number is in Table III.

Table III. Key to Investigators[a]

Table II Number	Investigator	Institution/Company
1	L. W. Burdett	Union Oil Company[b]
2	S. C. Blum	Exxon Corporation[b]
3	W. Barkley	Kettering Laboratory[b]
4	J. M. Holland	Oak Ridge National Laboratory
5	L. M. Holland	Los Alamos Scientific Laboratory
6	J. L. Epler	Oak Ridge National Laboratory
7	F. T. Hatch	Lawrence Livermore Laboratory
8	S. Zimmering	Brown University
9	M. Legator	University of Texas
10	H. P. Witschi	Oak Ridge National Laboratory
11	J. M. Giddings	Oak Ridge National Laboratory
12	D. L. Coffin	U. S. Environmental Protection Agency
13	J. N. Dumont	Oak Ridge National Laboratory
14	N. Richards	U. S. Environmental Protection Agency
15	W. H. Griest	Oak Ridge National Laboratory
16	B. R. Clark	Oak Ridge National Laboratory
17	L. Smith	Oak Ridge National Laboratory
18	W. Pepelco	U. S. Environmental Protection Agency
19	W. Birge	University of Kentucky

[a]See Table II
[b]Sponsored by the American Petroleum Institute

d) Simulated distillation to determine any changes in the boiling point distribution of the material which might be induced by loss of volatile constituents.

e) Elemental compositional scan by instrumental neutron activation analysis to detect inorganic contamination of the material by its container.

This battery of tests is being applied not only to the bulk samples in long-term storage in the drums and to aliquots for short-term distribution in borosilicate glassware, but also to aliquots taken in flint glassware and stored under various temperatures. This allows the stability of the materials to be studied under a variety of conditions to determine which is the optimum:

Amber Borosilicate = both room temperature and -30°C
Amber Flint = both room temperature and -30°C
Stainless Steel = 4°C

All of the above are under argon and are kept in the dark.

The results of the physical property measurements for the initial sampling at "time zero" (except where indicated) are shown in Table IV. Repetition of these measurements on the materials after one year of storage under various conditions shows little, if, any detectable change from the values listed (thus, they are not included here), with the possible exception of a 10 to 30 percent increase in the viscosity of the HDT Residue. Interestingly, the viscosity increased more in samples stored in borosilicate than in flint (a soft glass). This same container glass type effect has been observed in storage studies of No. 2 burner oils and diesel fuels, and was attributed not to an acceleration of fuel degradation by the borosilicate glass, but rather to an interaction of the soft glass with the fuel which slowed deterioriation. Although the soft glass container would appear to "stabilize" the material, it is probably an interaction with the material (which would partially contaminate it) which lends the apparent "stabilizing" effect upon some of the physical properties.

In the other stability tests, the infrared spectra recorded on a Fourier Transform IR spectrophotometer and the gas chromatograms generated on a 3 m x 3 mm glass column packed with a Dexsil 400 coated support did not show any detectable changes in the samples after one year of storage. This would suggest that no gross changes have occurred.

Stability testing will be continued on the samples at periodic intervals to monitor potential changes and follow up on the observations made after the first year of storage to determine if a trend is occurring.

CHEMICAL AND PHYSICAL CHARACTERIZATION

Part of the bioassay support function of the Facility is to perform chemical and physical measurements which would aid in interpretation of bioassay responses and endpoints. Part of this characterization is obtained in the baseline stability studies. Table IV indicates, for example, that the Crude Shale Oil, HDT Shale Oil, and HDT Residue are relatively viscous, high-boiling materials separate in properties from the other samples. It is interesting to note that the heavier materials are exhibiting mutagenicity and carcinogenicity in contrast to the renal effects of the refined products. Also, the data in Table IV suggest that the shale-derived jet and

Table IV. Physical Properties of the Crude, Process, and Refined Product Samples

Repository Sample No.	Description	Density (g/cm^3)	Viscosity (CS)	Boiling Points (°C) by Simulated Distillation		
				Initial	Average	Final
4601	Crude Shale Oil	0.917	50.5	20	358	514
4602	Hydrotreated Shale Oil	0.842	7.14	158	369	514
4603	Weathered Gasoline Feedstock	-	0.84	17	92	168
4604	JP-5 Precursor	-	1.58	124	201	271
4605	JP-8 Precursor	-	1.64	125	207	272
4606	DFM Precursor	0.835	3.74	152	278	330
4607	Hydrotreated Residue	0.866	20.6	255	401	513
4608	JP-5 Product	0.800	1.66	117	211	274
4609	JP-8 Product	-	1.60	113	206	273
4610	DFM Product	0.829	3.57	135	277	333
4612	Acid Sludge from DFM Treatment	-	130.	-	-	-
4614	Petroleum JP-5	0.805	2.16	60	225	287
4615	Petroleum JP-8	-	1.81	25	205	291
4616	Petroleum DFM	0.838	3.16	96	268	363

[a]Data from samples stored at -30°C in amber borosilicate glass.
[b]Measurements performed at 38°C; others in same column performed at 25°C.

diesel fuels are very similar to their petroleum equivalents. More detailed physical and chemical data on the shale-derived materials are available in a Union Oil Company report.[7]

Information on the general chemical class nature of the materials is generated in the course of their chemical class fractionation for bioassay. In this procedure, the materials are separated into acid/base fractions,[8] and the neutrals are further subdivided by gel chromatography[9] into various aliphatic and aromatic subfractions. Data averaged from one to four replicates from the bioassay preparation of this suite of material are listed in Table V. Although these data are derived from a preparative-scale procedure, and not an analytical method, they are useful for defining the general chemical nature of the organic constituents. The material with the greatest mutagnic activity, carcinogenicity, and toxicity is the Crude Shale Oil which contains the greatest percentage of acidic and basic constituents. However, all the materials are seen to be essentially neutral and paraffinic in character, particularly the finished fuels as is the objective of the refining. The relative amounts of the two aromatic fractions in the fuels and their precursors appears to parallel the volatility data in Table IV. The hydrotreating step appears to reduce acidic and basic constituents as might be expected from a partial removal of s- and n-heteroatomic functionality. Similarly, the acid/clay treatment appears to have reduced ether-soluble base constituents, going from the precursors to the products. Some difficulty was experienced in recovering fractions from the lighter, more volatile fuels and their precursors, and probably occurred in the solvent removal steps.

The elemental compositional scans of the stability study are not yet complete. However, the data generated in the baseline "time zero storge" studies a year ago have provided some information on the relative inorganic content of the materials. At least two distinct groupings of the materials can be made by inorganic content. The Crude Shale Oil, HDT Shale Oil, HDT Residue, and the Acid Sludge (the last possibly in a separate group, considering its sulfuric acid matrix) contain the highest concentrations of elements such as Na, K, Al, Mn, Cl, Cr, Fe, Zn, Br, Ag, Au, and La, whereas the jet and diesel fuel precursors and products contain 0.1 to 0.01 the concentrations of such elements. These differences would be expected from the refining.

IR spectra also were taken during the stability tests. Comparison of the IR spectra in Figure 1 for the Crude Shale Oil and HDT Shale Oil also shows the reduction of heteroatomic content by the hydrotreating process, and the similarity of the HDT Residue to the HDT Shale Oil. The spectra

Table V. Chemical Class Fraction Data for Crude, Process, and Refined Product Samples

Chemical Class Fraction	Percentage Composition[a]											
	No. 4601 Crude Shale Oil	No. 4602 HDT Shale Oil	No. 4603 W. Gas Feedstock	No. 4604 JP-5 Precursor	No. 4605 JP-8 Precursor	No. 4606 DFM Precursor	No. 4607 HDT Residue	No. 4608 JP-5 Product	No. 4609 JP-8 Product	No. 4610 DFM Product	No. 4614 Petroleum JP-5	No. 4616 Petroleum DFM
Volatile	0.3	1.2	36.1	1.9	2.1	0.7	0.6	3.8	5.5	0.6	2.3	1.0
Base (Ether Soluble)	3.7	1.3	0.16	2.25	4.1	2.2	0.8	0.1	0.2	0.02	0.3	0.02
Base (Insoluble)	1.0	0.4	0	0.6	0	0.7	0.7	-	0.2	-	-	-
Acid (Ether Soluble)	0.4	0.2	0.07	0.6	0.3	0.1	0.3	0.5	-	0.07	0.9	0.1
Acid (Insoluble)	1.1	0.3	0	0.5	0	0.5	0.6	-	0	-	-	-
Total Neutrals	93.3	92.2	27.1	89.1	86.1	105	99.4	90.8	90.3	97.9	93.9	97.5
Aliphatic	58.9	51.5	2.2	40.4	50.0	84.8	53.2	37.8	36.4	82.4	59.3	65.0
Mono/Diaromatic	26.4	21.0	0.05	5.2	5.7	12.3	34.2	4.5	5.2	7.2	4.1	17.9
Polyaromatic	2.8	4.1	0.6	0.09	0.2	0.6	10.0	0.1	0.1	0.1	0.1	1.3
Residue	2.1	4.6	-	2.4	1.3	1.8	5.1	1.1	1.8	2.1	2.0	1.3
Total Recovery[b]	97.7	95.3	63.4	48.1	63.7	102	102	47.5	49.4	92.5	68.2	86.7
Number of Replicates	2	4	1	2	2	2	2	2	1	1	2	1

[a]Percentage of original sample.
[b]Based on neutral fractions.

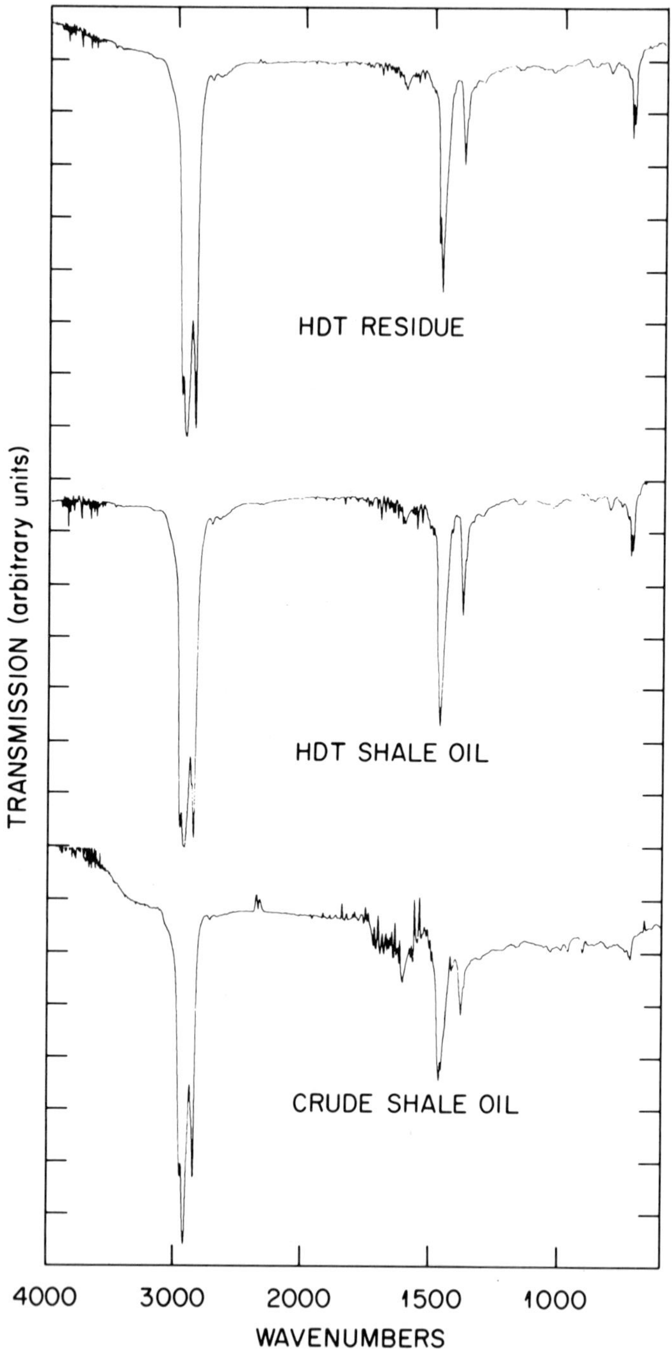

Figure 1. Infrared spectra of crude and HDT shale oils and HDT residue.

also illustrate the essentially hydrocarbon nature of these materials, with the spectra of the other materials being very similar to each other. No differences could be distinguished between the shale- and petroleum-derived products by IR.

Non-routine glass capillary column GC profiles of the diluted materials do show some interesting differences which correlate with the other data. Figure 2 shows the profiles for the Crude Shale Oil, HDT Shale Oil, and the HDT Residue. Hydrotreatment considerably increases the n-paraffin content (see also Table 6) and eliminates the baseline rise and "hump" in the gas chromatogram which would be associated with polar heteroatomic species which are not well resolved on a nonpolar stationary phase. The HDT Residue is seen to lack most of the C_{10}-C_{18} n-paraffins of the previous two samples (as is expected from its fractionation), but these more volatile species are recovered in the jet and diesel fuel products (see Figure 3 and Table VI).

Table VI. Measurements of Total n-Paraffins in Some Crude, Process, and Refined Samples

No.	Description	Conc., mg/g	Major Species
4601	Crude Shale Oil	27	C_{10}-C_{33}
4602	HDT Shale Oil	102	C_{10}-C_{34}
4607	HDT Residue	65	C_{18}-C_{34}
4608	JP-5 Product	197	C_{10}-C_{13}
4610	DFM Product	155	C_{10}-C_{18}
4606	DFM Precursor	158	C_{10}-C_{18}

Some small, minor differences in the chromatographable constituents of the shale- versus petroleum-derived products are visualized by this capillary GC technique. The glass capillary column GC profiles of diesel fuel marine derived from shale (upper chromatogram) and petroleum (lower chromatogram) in Figure 4 show some minor differences in the relative amounts of some constituents eluting near C_{17} and C_{18}. The petroleum-derived DFM also appears to contain a small percentage of higher paraffins than does the shale-derived DFM. Similar minor differences also were noted between the shale- and petroleum-derived jet fuels. The similarity in bioassay responses observed thus far for the petroleum- and shale-derived fuels would suggest, however, that these differences are negligible.

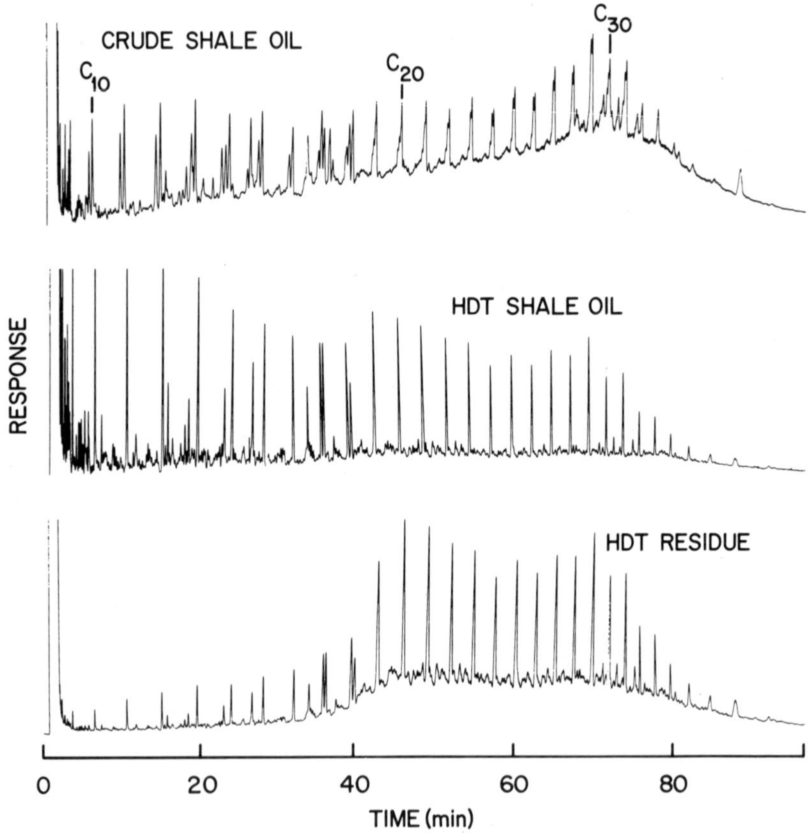

Figure 2. Capillary-column gas chromatograms of (A) crude shale oil, (B) HDT shale oil, and (C) HDT residue.

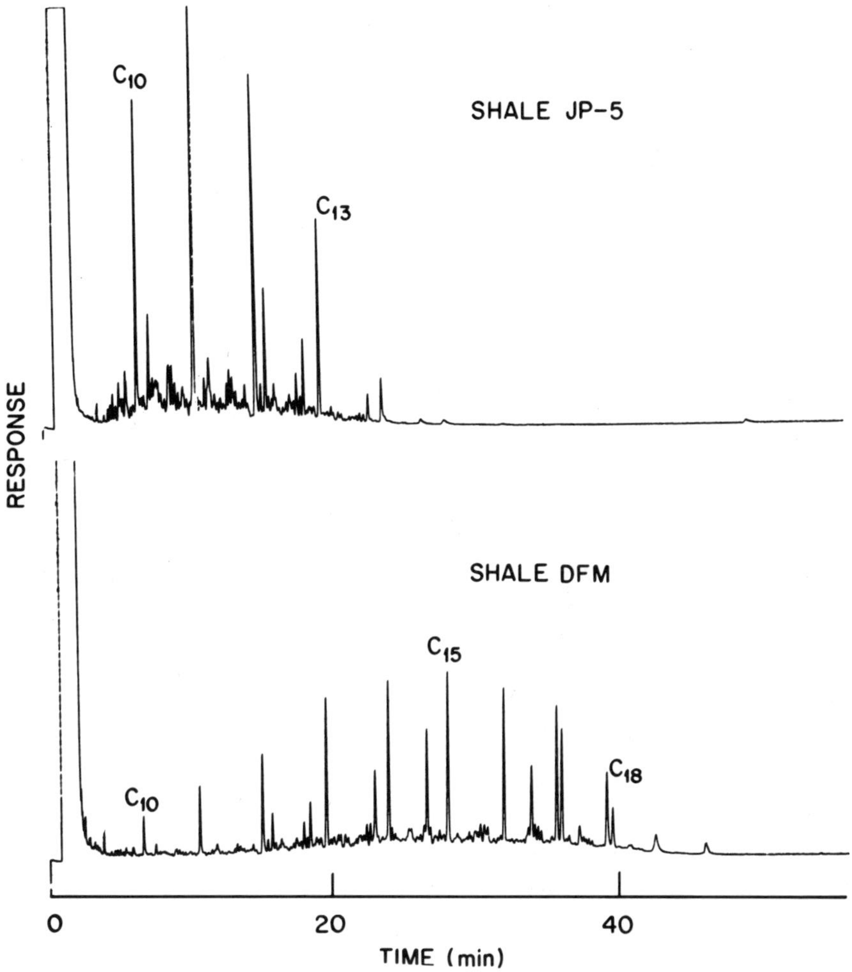

Figure 3. Capillary-Column gas chromatograms of (A) JP-5 product and (B) DFM product.

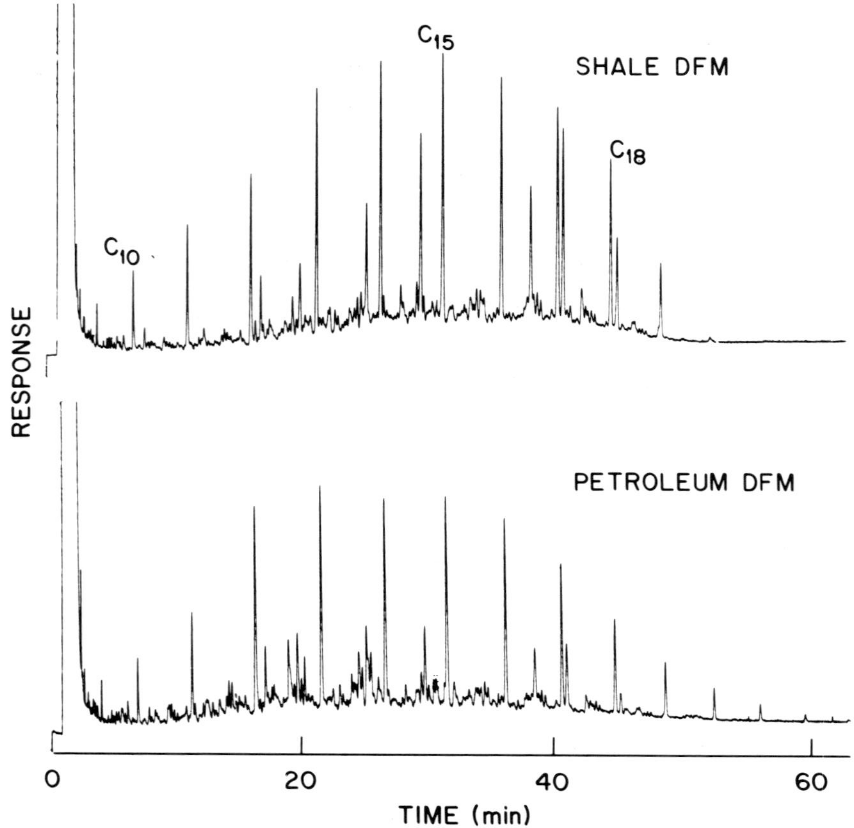

Figure 4. Capillary-Column gas chromatograms of DFM derived from (A) shale and (B) petroleum.

Specific bioactive constituents also are being measured in the samples to aid in bioassay response interpretation. Table VII contains data for measurements of benzo(a)pyrene in selected samples. It may be significant to note that although the benzo(a)pyrene content of the HDT Shale Oil and HDT Residue is equal to or greater than that of the Crude Shale Oil, the toxicity, mutagenicity, and carcinogenicity of the latter is greater than that of the former two. This result suggests that other classes of constituents, possibly those included in the relatively greater percentage of acidic and basic constituents of the Crude Shale Oil (Table 5), may be the determinant bioactive species. In contrast to these heavier materials, the jet and diesel fuels are seen to contain very little benzo(a)pyrene.

Table VII. Measurements of Benzo(a)pyrene in Some Crude, Process, and Refined Samples

No.	Description	Conc., μg/g
4601	Crude Shale Oil	11
4602	HDT Shale Oil	11
4607	HDT Residue	16
4608	JP-5 Product	N.D.[a]
4614	Petroleum JP-5	∿ 0.04
4606	DFM Precursor	0.03
4614	DFM Product	0.04

[a]ND = not detected.

The Facility will continue in its central role of sample management for this health effects investigation. The specific measurement of bioactive constituents will be extended, and the participation of other investigators in the study will be solicited.

REFERENCES

1. Personal communication, R. N. Hazlett, Combustion and Fuels Branch, Chemistry Division, Naval Research Laboratory, Washington, DC.

2. J. G. Christian, A. J. Chiantella, J. E. Johnson, and H. W. Carhart, "The Glass Effect in Distillate Fuel Stability," Ind. and Eng. Chem., 50, 1153-1156 (1958).

3. R. N. Hazlett, J. M. Hall, and J. C. Burnett, "Aging Behavior of Crude Shale Oil," NRL Memorandum Report 3844, Naval Research Laboratory, Washington, DC August 1978.

4. D. W. Brinkman, M. L. Whisman, and J. N. Bowden, "Stability Characteristics of Hydrocarbon Fuels from Alternative Sources," BETC/RI-78/23, Bartlesville Energy Technology Center, Bartlesville, OK, March 1979.

5. F. S. Karn, F. R. Brown, and R. A. Friedel, "Aging Studies of the Synthoil Product. Part I: Viscosity Changes," PERC/TPR-76/1, Pittsburgh Energy Technology Center, Pittsburgh, PA, May 1976.

6. F. S. Karn and F. R. Brown, "Aging Studies of the Synthoil Product. Part II: Effects of Time, Temperature, Atmosphere," PERC/TPR-76/2, Pittsburgh Energy Technology Center, Pittsburgh, PA, September 1976.

7. M. R. Winward and L. W. Burdett, "Comprehensive Analyses of Shale Oil Products: Second Study," Science and Technology Division Projct Report No. 79-39, Draft copy submitted to the American Petroleum Institue by the Union Oil Company of California, November 27, 1979.

8. I. B. Rubin, M. R. Guerin, A. A. Hardigree, and J. L. Epler, "Fractionation of Synthetic Crude Oils from Coal for Biological Testing," Env. Res., 12, 358-365 (1976).

9. A. R. Jones, M. R. Guerin, and B. R. Clark, "Preparative-Scale Liquid Chromatographic Fractionation of Crude Oils from Coal and Shale," Anal. Chem., 49, 1766-1771 (1977).

CHAPTER 5

ANALYSIS OF SHALE OIL LIQUIDS FOR POLYNUCLEAR AROMATIC HYDROCARBONS AND THEIR N-HETEROCYCLIC ANALOGS

W. K. Robbins, S. C. Blum, Exxon Research and Engineering, Linden, New Jersey, 07036

INTRODUCTION

Renewed interest in shale oil liquids has prompted a number of health related studies including biological testing of various shale oil fractions.[1,2,3] Mutagenic activity has been isolated largely into two fractions: the non-basic aromatics which include the polynuclear aromatics (PNAs) and the basic aromatics could include both N-heterocyclics or amino aromatics (N-PNAs). Few specific compounds have been measured in either fraction. The American Petroleum Institute Project SPS-5 was undertaken with four objectives:

(1) To develop a separation scheme for the shale oils which would yield an aromatics rich fraction that could be analyzed for PNAs by the GC/UV procedure.[4]

(2) To adapt the GC/UV procedure for specific N-PNAs (carbazole, acridine and their benzo analogs).

(3) To apply the GC/UV procedure for specific N-PNAs to a N-rich fraction isolated in the separation scheme.

(4) To qualitatively characterize other N-heterocyclics in the isolated fractions.

This paper reports the analysis of 23 PNAs and 14 selected N-PNAs in six shale oil liquids including five from the SOHIO-Paraho suite.

DEVELOPMENT OF A SHALE OIL SEPARATION SCHEME FOR GC/UV ANALYSES

The GC/UV procedure relies on the preparation of a PNA-rich fraction prior to the final isolation of individual compounds by prep-GC. Experience with a wide range of petroleum matrices suggested that a combination of chromatography on partially deactivated silica gel (e.g. silicic acid) and liquid-liquid extraction with N-methyl pyrrolidone (NMP) could provide such a PNA-rich fraction suitable for GC/UV analysis. Such a procedure has been developed (Figure 1).

Deactivated silica gels (e.g. silicic acids) have been used previously for the separation of aromatics and polar aromatics from complex matrices.[5,6,7] To determine the most appropriate conditions for isolating PNAs and N-PNAs from shale oils, the distribution coefficients (K_{Ds}) of a number of model compounds were measured for several absorbent-eluent combinations.[8] Ten percent water deactivated silicic acid was found to be the most useful adsorbent for the range of polarity represented by the PNAs and N-PNAs. These K_{Ds} were used to calculate the elution volumes given in Table I which demonstrate that a 60 g column may be used to obtain reasonably well-resolved fractions which contain the saturates, aromatics (including PNAs), non-basic N-PNAs, basic N-PNAs, and highly polar components. The fractionation of other compound classes on this column was not studied here.

Previous experience with petroleum matrices has shown that the NMP-Acid partitioning is a useful technique for the isolation of PNAs from other UV-absorbing materials in complex matrices.[9] Distribution coefficient data for model compounds indicate that the N-PNAs are extracted efficiently into the NMP Acid (Table II). Back extraction was poor under the PNA conditions; but when the NMP-Acid extract is diluted with an eight-fold excess of water and neutralized to pH > 9, nearly quantitative recoveries are attained (Table III). The two N-PNA fractions are combined in the shale oil analyses for three reasons: (1) Each fraction is to be extracted before GC/UV analysis, (2) the split between basic and non-basic N-heterocyclics is complicated by the number of alkyl derivatives and structural isomers which overlap one another and (3) only one ^{14}C-tracer is presently available for determining recoveries of N-heterocyclics.

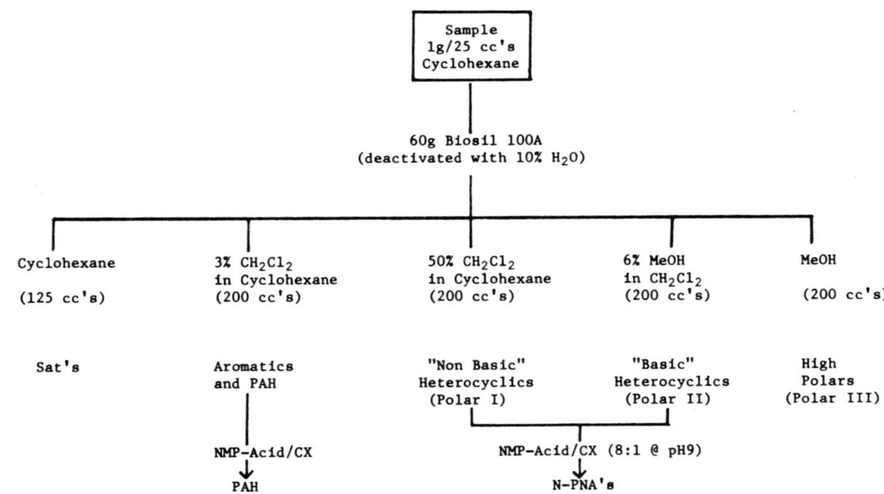

Figure 1. Separation scheme for shale liquids.

Table I. Calculated Elution Volumes for Selected PAH and N-PNA's on 10% H_2O Deactivated Silicic Acid (60 g column, void volume = 60 cc)

Compound	Cyclohexane	3% CH_2Cl_2 in Cyclohexane	50% CH_2Cl_2 in Cyclohexane	6% CH_3OH in CH_2Cl_2	CH_3OH
Phenanthrene	168	78	V[b]	V	V
Benz(a)anthracene	228	138	V	V	V
Benzo(a)pyrene	240	138	V	V	V
Dibenz(a,h)anthracene	354	168	V	V	V
Carbazole	C[a]	C	142	V	V
Benzo(c)carbazole	C	C	150	V	V
7H-Dibenzo(c,g)carbazole	C	C	126	V	V
Dibenz(a,h)acridine	C	C	420	V	V
Benz(c)acridine	C	C	270	V	V
Acridine	C	C	C	180	V
Quinoline	C	C	C	102	V

[a] C = quantitatively retained on the column.
[b] V = eluted in the first void volume of solvent.

Table II. Partition Coefficients for Extraction of N-PNA by 4:1 NMP-Acid from Cyclohexane

Compound	K_D	% Extracted[a]
Acridine	100	99
Carbazole	32	99
Phenanthridine	11	99
Benz(c)acridine	5.0	98
11-H Benzo(a)carbazole	33	99
2-Azafluoranthene	12	99
Azapyrene	15	99
1,10-Dimethylbenz(a)acridine	1.4	80
Dibenz(a)acridine	8.8	99
Dibenz(a,j)acridine	24	99

[a]Calculated from theory for a 2:1 ratio of solvent to extractant and three extractions.

TABLE III. Partition Coefficients for Back Extraction of N-PNA by Cyclohexane from NMP:H_2O (1:8) at pH > 9

Compound	K_DB[a]	Theoretical % Recovered[b]
Acridine	12	87
Carbazole	17	92
Phenanthridine	17	92
Benz(c)acridine	>100	100
Benzo(a)carbazole	>100	100
2-Azafluoranthene	23	95
Azapyrene	23	95
1,10-Dimethylbenz(a)acridine	>100	100
Dibenz(a,h)acridine	>100	100
Dibenz(a,j)acridine	>100	100

[a]K_DB = partition coefficient for cyclohexane as extractant, i.e., back extraction.
[b]Calculated from theory for two cyclohexane extractions of 50 ml on a 350 ml aqueous-NMP phase.

SEPARATION OF SHALE OIL LIQUIDS

The mass distributions for the shale oil liquids on the silicic acid column show that the raw shale oils contain >20% of their mass in the N-rich fractions while the processed oils contain < 10% (Table IV). The "unaccounted" mass is due to either volatilization of front ends or column hold-up depending on the sample. For example, N_2 stripping of a bulk aliquot of the *in situ* shale oil resulted in a 16.4% loss. C, H, and N concentrations were determined for residues which contained sufficient material. The data shown in Table V confirm that the column permitted a good separation of the N-containing components from the saturates and aromatics. The lack of an elemental material balance for each polar fraction suggests the presence of S or O.

When the silicic acid column preceded the NMP extraction, the procedure obtained good recoveries of the ^{14}C-tracers. For the PNAs, the recoveries of tracers from the silicic acid column were nearly quantitative (Table VI). The subsequent NMP extraction yielded about 90% recovery while reducing background UV absorption ten fold. With the exception of the hydro-resid sample, the column generally provided good yields of N-PNA based on the ^{14}C activity (Table VII). The subsequent NMP extraction (modified for N-PNA) produced 90% recovery of ^{14}C-activity while reducing the UV background at 300 nm 100-fold (Table VIII).

When the reverse sequence (NMP partitioning followed by silicic acid chromatography) was applied to raw shale oils, emulsions were encountered. Pre-extraction of the sample with acid and/or base failed to alleviate this problem. Because the emulsions caused low ^{14}C-recoveries (Table VI), this approach was discontinued.

PNA ANALYSES OF THE SHALE LIQUIDS

The results of the GC/UV analysis for the five Paraho related shale oils are given in Table IX; duplicate analyses of the *in situ* (Rock Springs, Site 9) are given in Table X. The duplicate results illustrate the characteristic repeatability of the GC/UV procedure except for fluorene. In the latter case, the second result is believed to be low due to inadequate cooling of the trap for this, the most volatile PNA that is measured by the technique. These results are similar to those obtained on comparable petroleum fractions.[10]

Table IV. Gravimetric Distribution of Shale Oil Samples

Class	% Recovery					
	4601.3 Crude Shale Oil	4602.3 Hydro. Shale Oil	4607.3 Hydro. Resid	4606.3 DFM Precursor	4610 DFM Product	In Situ Shale Oil
Saturates	31.7	66.5	75.7	75.7	75.4	32.2
Aromatics	11.2	9.8	14.0	6.4	7.5	14.5
N-Heterocyclics (Non-Basic)	15.1	2.5	4.0	1.1	0.8	8.1
N-Heterocyclics (Basic)	17.9	2.9	3.8	2.8	0.4	16.3
Polars	15.8	0.9	1.0	0.4	0.2	11.0
Loss	8.3	17.4	1.5	13.5	15.7	17.9

Table V. Elemental Analyses of Shale Oil Fractions

Class		4601.3 Crude Shale Oil	4602.3 Hydro. Shale Oil	4607.3 Hydro. Resid	4606.3 DFM Precursor	4610 DFM Product	In Situ Shale Oil
	%C	86.22	86.52	86.79	85.49	86.53	85.31
Cut 1	%H	13.44	14.07	14.04	13.98	13.92	14.24
(Sats)	%N	0.3	<0.3	<0.3	<0.3	<0.3	<0.3
	%C	85.96	90.72	90.32	90.28	89.72	86.44
Cut 2	%H	9.94	9.38	9.67	10.06	10.35	8.36
(Aromatics)	%N	0.77	0.3	<0.3	<0.3	<0.3	<0.3
Cut 3	%C	83.98	85.08	86.36	a	a	80.60
(Non-Basic)	%H	10.29	8.53	8.91	a	a	9.86
N-Heterocyclics	%N	2.74	4.89	4.40	a	a	3.5
Cut 4	%C	82.18	83.84	84.98	a	a	80.87
(Basic)	%H	10.28	10.99	11.29	a	a	9.57
N-Heterocyclics		4.09	4.27	3.65	a	a	b
Cut 5	%C	67.43	76.87	77.39	a	a	73.79
	%H	10.15	10.13	10.35	a	a	9.20
Polars	%N	3.98	4.41	3.98	a	a	b

[a] Insufficient sample for C, H, N
[b] N values not determined

Table VI. Recovery of ^{14}C-Activity in PNA Fractions in Separation Steps

Code	Description	% After Silicic Acid Column	% After Column and NMP
4601.3	Crude Shale Oil	--	48[a] 104
4602.3	Hydro Shale Oil	--	49[a]
4606.3	DFM Precursor	93.6	80
4607.3	Hydro Resid	124	94
4610.3	DFM Product	96	92
--	In Situ Shale Oil	102 107	95 85

[a]Separated by reverse sequence, i.e., NMP then column.

Table VII. Recovery of ^{14}C-Activity in N-PNA Clean-Up

Code	Sample	% Recovered After Silicic Acid Column	Approximate % Recovered After Column and NMP Partition[a]
4601.3	Crude Shale Oil	97	89
4602.3	Hydro Shale Oil	79	64
4606.3	DFM Precursor	91	89
4607.3	Hydro Resid	31	26
4610.3	DFM Product	100	92
--	In Situ Shale Oil	122	115
	In Situ Shale Oil	126	108

[a]Samples were reduced to approximately 25 cc in beakers and ∿ 1 cc removed for radio assay.

Table VIII. Approximate Recoveries for NMP-Acid Partitioning of N-PNA Column Fractions[a]

		% of Column Activity	% of UV Absorvance at 300 nm
4601.3	Crude Shale Oil	92	1.1
4602.3	Hydro Shale Oil	82	0.9
4606.3	DFM Precursor	98	0.3
4607.3	Hydro Resid	85	1.3
4610.3	DFM Product	92	1.5
--	In Situ	95 89	0.7 -

[a]Samples were reduced to approximately 25 cc in beakers - then the UV spectra were recorded and ∿ 1 cc aliquot was removed for radio-assay.

Table IX. PNA Measured by GC/UV in Paraho Shale Liquids

	4601.3 Crude Shale Oil	4602.3[a] Hydrotreated Shale Oil	4607.3 Hydrotreated Residue	4606.3 DFM Precursor	4610 DFM Product
	μg/g	μg/g	μg/g	μg/g	μg/g
Fluorene	< 29	28	14	208	124
Phenanthrene	77	133	124	193	281
Anthracene	11	< 10	< 10	< 10	< 10
Fluoranthene	4.7	23	109	< 0.5	< 1
Pyrene	2.6	146	287	0.5	1
Benz(a)anthracene	9	12	10	0.5	1
Chyrsene	< 4.6	16	16	0.5	1
Triphenylene	< 6.5	2.0	1.7	0.5	1
Me-B(a)A	< 4.6	23	58	0.5	1
DMe/Et-B(a)A	< 4.6	5.0	48	0.5	1
Benzo(g,h,i)fluoranthene	< 10	125	276	0.5	1
Benzo(b)fluoranthene	< 10	3.8	< 12	0.5	1
Benzo(j)fluoranthene	< 10	< 5.1	< 11	0.5	1
Benzo(k)fluoranthene	< 1.5	1.0	1.6	0.5	1
Perylene	< 1.1	2.5	3.7	0.5	1
Benzo(e)pyrene	< 5.5	40	54	0.5	1
Benzo(a)pyrene	2.0	10	14	0.5	1
Methyl B(e)P	< 1.7	44	39	0.5	1
Methyl B(a)P	< 5.7	12	11	0.5	1
1,2,3 Indeno(c,d)pyrene	< 0.9	0.5	< 1.1	0.5	1
Dibenz(a,h)anthracene	< 3.2	1.1	< 3.9	0.5	1
Benzo(g,h,i)perylene	5.0	77	83	0.5	1
Coronene	< 1.7	8.8	11	< 0.5	< 1

[a]Analyzed by Method A, all others analyzed by Method B.

TABLE X. PNAs MEASURED BY GC/UV IN IN-SITU SHALE OIL (ROCK SPRINGS, WYOMING - SITE 9)

	μg/g	
	1st Analysis	2nd Analysis
Fluorene	564	152
Phenanthrene	2698	2439
Anthracene	497	449
Fluoranthene	361	323
Pyrene	462	541
Benz(a)anthracene	150	124
Chrysene	164	128
Triphenylene	27	35
Me-B(a)A	52	45
DiMe/Et B(a)A	29	26
Benzo(g,h,i)fluoranthene	<67	<18
Benzo(b)fluoranthene	<28	22
Benzo(j)fluoranthene	<11	<15
Benzo(k)fluoranthene	11	15
Perylene	6.7	9.2
Benzo(e)pyrene	70	76
Benzo(a)pyrene	67	60
Me-B(e)P	34	10
Me-B(a)P	9	16
1,2,3 Indeno(c,d)pyrene	1.3	1.9
Dibenz(a,h)anthracene	5.2	4.9
Benzo(g,h,i)perylene	35	37
Coronene	1.6	2.3

The results for the DFM fractions indicate the presence of only the smallest 3 ring PNAs by the GC/UV technique. The GC trace contained peaks with retention times up to approximately the retention time of pyrene, but none of these peaks exhibited recognizable spectra aside from the reported results. At longer retention times only the peaks for the tracer compounds were observed.

In contrast, the chromatograms for both the hydrotreated oils (4602.3 and 4607.3) exhibited numerous broad peaks above a substantial background. The retention times of some of these peaks were greater than coronene. The data for benz(a)anthracene and its methyl derivatives in both samples suggest the alkyl-substituted compounds constitute a significant portion of the PNA content. This is further supported by the observation that several of the GC/UV traps following the parent pyrene peak exhibit pyrene-like spectra. In the absence of alkyl pyrene absorptivity values, the concentrations of these derivatives can only be estimated, but are equivalent to or greater than the parent level. This is consistent with results reported by Jones and co-workers, who reported that the PNAs in an unidentified shale oil exhibit high substitution with the maximum abundance around C_5 or C_6.[11] Less characteristic alkyl derivatives undoubtedly contribute a large portion of the background material upon which the parent peaks are observed. These compounds also account for much of UV background found in the isolated fractions. However, in these hydrotreated samples, the parent or the methyl substituted components are sufficiently enriched to allow detection of nearly all the 23 compounds in the analytical array.

The gas chromatograms for the two raw shale oils also exhibit a high general background. However, while the *in situ* oil chromatogram is characterized by a number of large peaks which correspond to the reported PNA, the crude Paraho shale oil exhibits few major peaks beyond the 3-ring region. The large quantities of unresolved material contribute a strong UV tail which lacks any characteristic peaks in the solutions obtained from the GC traps. The combination of a high mass of unresolved UV absorbing material resulted in larger than normal "less-than" values for the analysis of PNA in the Paraho shale oil. Nevertheless, it is obvious that the two raw shale oils contain different levels of PNAs.

Table XI. GC/UV Characteristics for Heterocyclic Compounds

Compound	Retention Time[a]	Wavelength nm	Baseline nm	Absorptivity[b] (ml/μg-cm)
Carbazole	5.7	291	288-297	0.0947
Benzo(h)quinoline	5.7	346	341-350	0.0192
Acridine	6.0	249	224-254	0.9929
Phenanthridine	6.1	343	350-332	0.0101
Benzo(f)quinoline	6.6	345	340-350	0.0126
2-Azafluoranthene	10.3	365	354-380	0.0300
2-Azapyrene	11.7	275	254-283	0.1903
Benz(c)acridine	16.2	383	370-392	0.0792
Benz(a)acridine	17.6	382	369-390	0.0588
11-H Benzo(a)carbazole	18.3	351	346-355	0.0297
1,10 Dimethyl Benz(a)acridine	21.8	387	376-395	0.0444
8,12 Dimethyl Benz(a)acridine	23.2	389	377-400	0.0409
2,10 Dimethyl Benz(a)acridine	23.4	386	375-394	0.0416
13-H Dibenzo(a,i)carbazole	23.7	346	337-351	0.0343
Dibenz(a,h)acridine	29.9	394	380-402	0.0974
Dibenz(a,j)acridine	30.4	393	379-400	0.0936
7H Dibenzo(c,g)carbazole	31.6	362	351-371	0.1815

[a]GC conditions: Column: 10 ft. x 1/8 in., 2% SE-30 on Chrom G.
Flow: 40 ml/min
Prog.: In @ 175°C, prog. @ 4°C/min to 300°C and hold.

[b]In cyclohexane

EXTENSION OF THE GC/UV PROCEDURE TO N-PNAs

The extension of the GC/UV procedure to N-PNAs was straightforward. Seventeen N-PNAs representing 3, 4 and 5 ring structures were obtained along with ^{14}C-labeled dibenz(a,h)acridine. These compounds included most available acridines, carbazoles, and their benzoanalogs as well as an azafluoranthene and an azapyrene. Each class of N-PNA contains many more isomers than the corresponding PNA. For example, only two PNAs correspond to mass 178 (phenanthrene and anthracene) while there are eight N-PNAs at the analogous mass of 179 (Figure 2). Although not all of these eight N-PNAs were obtained as reference compounds, those available provide a good indication of the GC and UV characteristics of this class of 3-ring compounds (Table XI).

The GC retention times of the N-PNA are similar to those of the corresponding PNAs as would be expected. Separations on non-polar silicones such as SE-30 follow the boiling points which are very similar for the N-PNA and PNA compounds. All of the compounds have symmetric peaks with the specified stainless steel columns, indicating no hold-up in the GC. Recovery of the N-PNA from the stainless steel traps, however, required a solvent more polar than cyclohexane (Table XII). In most of the N-PNA analyses the traps were washed with methylene chloride and solvent exchanged into cyclohexane. The solvent exchange was required because the spectra of the compounds in methylene chloride lacked the structural detail necessary for identification and quantification. Because the solvent exchange is time consuming, alternate solvents were considered. During the course of this project, methyl-t-butyl ether (MTBE), which has solvency characteristics similar to methylene chloride, became commercially available in high purity. Since the UV spectra of the N-PNA and PNA are identical in cyclohexane and MTBE, the latter solvent was adopted for trap washing. The MTBE was used for two of the shale liquid analyses.

Table XII. Trapping Efficiencies and % Recoveries of Representative N-PNA

	Cyclohexane Eluent	Methylene Chloride Eluent
Acridine	--	115
Benz(c)acridine	83	111,97
Dibenz(a,h)acridine	--	105
7H-Dibenz(c,g)carbazole	65,61	93

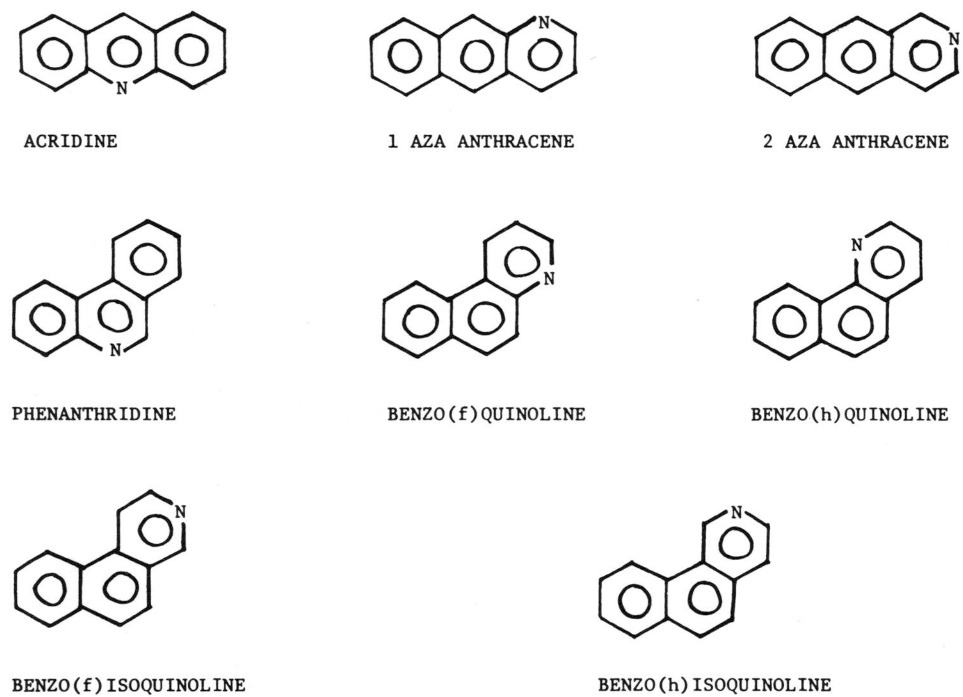

Figure 2. N-analogs of anthracene and phenanthrene.

N-PNA ANALYSES OF SHALE OILS

The results of the GC/UV analysis for N-PNA in the six shale oil liquids are given in Table XIII. The detection limits vary for individual compounds due to differences in the absorptivities; detection limits vary between analyses due to differences in the aliquot fraction injected into the chromatograph. In all the analyses, a majority of detection limits were 2 μg/g or lower (Table XIV). Only measured N-PNA are reported in Table XIII. (The three dimethylbenz(a)-acridines included in the reference compound list have been deleted since they represent only a small fraction of the possible isomers.)

Detailed study of the UV spectra for the GC trapped components indicated that several of the samples contained alkyl substituted N-PNA. Since reference compounds for such isomers were not available, the concentrations of these materials have been estimated from the absorptivity of the parent compounds Table XV). Even with these estimated components considered, the six samples appear to contain only a few discrete N-PNAs, most notably carbazoles and benzocarbazoles. No N-PNAs were detected in the crude shale oil (4601.3) at the levels studied.

In contrast, the gas chromatograms generated in the isolation procedure indicate a large number of significant well resolved components for all the samples except the crude shale oil (4601.3). The latter sample exhibited small peaks above a large background similar to that observed for the PNA. Although some of the trapped materials exhibited potentially useful spectra, most of them yielded non-descript curves with insufficient detail for identification or quantification.

QUALITATIVE CHARACTERIZATION OF N-HETEROCYCLICS

The GC/UV procedure has allowed the analysis of the selected N-PNA compounds. The detection limits were adequate, but because the introduction of a N into each PNA structure greatly increases the number of isomers, these surrogates, suggested from air pollution studies, may be incomplete for shale oil analysis.[12] To obtain a better understanding of the materials involved, the residues not consumed in the GC/UV analyses were analyzed by low voltage, high resolution mass spectrometry (LV/HR/MS). If necessary, a drop of toluene was added to the sample and taken up in a 5 microliter dipper. This sample was charged into a MS-902S

Table XIII. N-PNAs Measured by GC/UV

	4602.3 Hydro. Shale Oil	4607.3 Hydro. Residue	4606.3 DFM Precursor	4610.3 DFM Product	In Situ Shale Oil
	(μg/g)	(μg/g)	(μg/g)	(μg/g)	(μg/g)
<u>3 Ring</u>					
Carbazole	2.0	4.9	31.7	4.9	154
Benzo(h)quinoline	--	--	--	--	56
<u>4-Ring</u>					
Azapyrene	--	13	--	--	--
Benzocarbazole	3.0	22	--	--	71
Benzacridine	--	--	--	--	17
<u>5 Ring</u>					
Dibenz(a,h)acridine	--	--	--	--	6

Table XIV. Detection Limits for N-PNAs in the GC/UV Procedure for Shale Oil

	4601.3 Crude Shale Oil	4602.3 Hydro. Shale Oil	4607.3 Hydro. Resid	4606.3 DFM Precursor	4610.3 DFM Product	In Situ
	(μg/g)					
<u>3 Ring</u>						
Carbazole	0.9	1.0	0.5	0.3	0.7	2.3
Benzo(h)quinoline	5	5.2	17	1.7	0.1	12.0
Acridine	0.1	0.1	0.5	0.5	0.1	0.2
Phenanthridine	0.5	0.5	3.7	0.2	3.4	1.3
Benzo(f)quinoline	3.7	4.0	13	1.3	5.3	9.2
<u>4 Ring</u>						
2-Azafluoranthene	3.0	3.5	11	1.1	2.2	7.7
2-Azapyrene	0.3	0.4	1.1	0.1	0.2	0.8
11-H Benzo(a)carbazole	3.4	3.4	11	1.1	2.2	7.7
Benz(c)acridine	0.9	1.3	4.1	0.4	0.9	2.9
Benz(a)acridine	1.6	1.7	5.5	0.5	1.0	3.9
<u>5 Ring</u>						
13-H Dibenz(a,i)carbazole	2.8	2.9	9.4	0.9	2.0	6.7
Dibenz(a,h)acridine	1.0	1.0	3.5	0.5	0.7	2.4
Dibenz(a,j)acridine	1.0	1.0	3.3	0.4	0.7	2.4
7-H Dibenzo(c,g)carbazole	0.5	0.6	1.8	0.2	0.3	1.2

Table XV. Alkyl N-PNAs Estimated in Shale Liquids From GC/UV Data

	($\mu g/g$)					
	4601.3 Crude Shale Oil	4602.3 Hydro. Shale Oil	4607.3 Hydro. Resid	4606.3 DFM Precursor	4610.3 DFM Product	In Situ
3 Ring						
C_n Carbazoles	--	6	18	10	5	702
C_n Benzoquinoline	--		43			
4 Ring						
C_n Azapyrenes	--	11	128			
C_n Benzocarbazoles	--					112
C_n Benzacridines	--					13

Table XVI. Distribution of N-Compounds in LV/HR Mass Spec Analyses

	4601.3 Crude Shale Oil	4602.3 Hydro. Shale Oil	4607.3 Hydro. Resid	4606.3 DFM Precursor	4610.3 DFM Product	In Situ Shale Oil
A. Aromatic Ring No.	%	%	%	%	%	%
1	15	21	16	21	4	4
2	31	30	30	61	48	33
3	35	31	32	17	50	46
4	12	15	17	1		13
5	6	3	2	-		3
B. Ring Substitution						
C_0	1.3	1.3	4.3	4.2	22	12
C_3[a]	21	12	17	25	0[c]	21
C_4[b]	19	16	16	19	0	10

[a]Corresponds to either $3CH_3$'s etc., or a 3 carbon sat'd bridge for example tetrahydrocarbazole or C_3 Quinoline.
[b]Corresponds to either 4 CH_3's etc., or to a 4 carbon sat'd ring; for example C_4 acridine, C_2 tetrahydroazapyrene, or hexahydrobenzacridine.
[c]All the components in DFM product were C_0, C_1, or C_2.

high resolution mass spectrometer. An additional four microliters of a mixture of halogenated aromatics was added as an internal standard. The nitrogen containing components were then sorted by Z-series and standard sensitivity corrections ere made for each homolog within the Z-series. Since only nitrogen-containing masses were detected in the basis set, the data was normalized to 100% (wt/wt) to allow the compound type distributions to be examined.

Because the data demonstrate that all the ionized material (i.e., all the aromatics) contained some nitrogen, the Z-number matrices indicate that:

(1) A majority of the N-PNA fraction consisted of compounds with 3 or fewer aromatic rings (Table XVI).

(2) Most classes of compounds exhibit large components with C_1 to C_6 substitution which maximize around C_3 or C_4 for the Paraho shale liquids, the <u>in situ</u> oil exhibits more C_0-C_2 material suggesting a higher temperature pyrolysis (Table XVII).

(3) Of the <u>nominal</u> structures corresponding to the Z--numbers, carbazoles appear to be the dominant N-PNA class of compounds (Table XIII).

The MS methodology is well suited for grouping related compounds into classes on the basis of unsaturation and alkyl derivatization, but it does not distinguish between structural isomers of the same mass. For example, azafluorenes, aminofluorenes, and dihydroacridines, etc., are all grouped with alkyl carbazoles at the same isomer mass. Although MS analyses of the shale liquids showed masses that would include the alkyl carbazoles, the GC/UV method did not confirm their presence in all cases, most notably in the crude shale oil. Such GC/UV results would suggest that the MS is responding to other structural isomers for which reference spectra are not readily available. These results are consistent with those of earlier workers who found evidence for partially hydrogenated multi-ring structures.[13,14]

SUMMARY

Six shale oil liquids have been analyzed for PNAs and selected N-PNAs. These samples were found to contain PNAs at levels similar to those in comparable petroleum fractions. Few of the selected acridines, carbazoles or benzo analogs which could be measured by the GC/UV procedure were detected.

Table XVII. Distribution of N-Compounds in LV/HRMS - Ring Substitution

	%					
	4601.3 Crude Shale OIl	4602.3 Hydrotreated Shale Oil	4607.3 Hydrotreated Residue	4606.3 DFM Precursor	4610.3 DFM Product	In Situ Shale Oil
Ring Substitutions						
C_0	1.3	1.3	4.3	30	25	16
C_1	3.8	6.2	7.2	28	60	24
C_2	18	20	15	12	15	26
C_3	21	12	17	13	--	18
C_4	19	16	16	20	--	8
C_5+	15	12	9	16	--	6

Table XVIII. Compound Types by MS Z-Number

Z	Nominal Structures	4601.3 Crude Shale Oil	4602.3 Hydro. Shale Oil	4607.3 Hydro. Residue	4606.3 DFM Precursor	4610.3 DFM Product	In Situ Shale
3	Pyrroles	0.9	0	1	0	4	.3
5	Pyridines	13	21	15	21	0	4
7	Azaindanes	5	6	9	8	0	2
9	Indoles	16	8	5	45	0	24
11	Quinolines	6	8	8	6	31	5
13	Phenyl Pyridines	4	7	7	2	16	3
15	Carbazoles	19	20	14	11	48	26
17	Acridines	9	7	14	4	--	15
19	Benzo(d,e,f)carbazoles	6	4	4	0	--	4
21	Azapyrenes, Benzcarbazoles	8	11	13	.5	--	10
23	Benzacridines	3	3	4	0	--	3
27	Dibenzcarbazoles	2	0.9	1	0	--	1.3
29	Dibenzacridines	3	0.4	0.2	0	--	0.2

However, elemental analysis and LV/HR/MS demonstrate that numerous N-containing compounds are isolated in the fractions which would contain N-PNAs. Qualitatively, the LV/HR/MS data suggest that the bulk of the N is associated with highly substituted or partially hydrogenated multi-ring heterocyclics.

ACKNOWLEDGEMENTS

The authors express thanks to L. Crisson who performed most of the experimental work, to F. C. McElroy for the development of the column separation technique, and to J. M. Kelliher who carried out the MS analyses.

REFERENCES

1. J. L. Epler and M. R. Guerin, "Mutagenic Components of Alternate Energy Sources," Proced. 71st Air Pollution Control Assoc. Annual Meeting, Air Pollution Control Association, paper 28.23.5.

2. R. A. Pelroy and M. R. Petersen, "Mutagenicity of Shale Oil Components," Environ. Sci. Res. 15, 463-475, 1979.

3. B. R. Clark, C. H. Ho, W. H. Griest, and M. R. Guerin, "Biologically Important Compounds in Synfuels Process," Preprints, ACS Div. of Fuel Chemistry 25(2), 1, 1980.

4. R. A. Brown and T. D. Searl, "Analysis of Polynuclear Aromatic Hydrocarbons," Chapter 16 in Chromatography in Petroleum Analysis, K. H. Altgelt and T. H. Gouw, pp. 367-394 (1979).

5. L. R. Snyder and B. E. Buell, "Nitrogen and Oxygen Compound Types in Petroleum," Anal. Chem. 40, 1295 (1968).

6. L. R. Snyder, "The Nitrogen and Oxygen Compounds in Petroleum," Preprints, ACS Div. of Petroleum Chem. 15, C-43 to C-62 (1970).

7. M. Selucky, T. Ruo, Y. Chu, and O. P. Strausz, "Chromatographic Studies in Oil Sand Bitumens," Analytical Chemistry of Liquid Fuel Sources, Amer. Chem. Society, pp. 117-127 (1977).

8. F. C. McElroy and W. K. Robbins, paper in preparation.

9. W. K. Robbins, "Solvent Extraction of Polynuclear Aromatic Hydrocarbons," presented at the 4th International Symposium on Polynuclear Aromatic Hydrocarbons, Columbus Ohio, October 1979.

10. R. J. Pancirov and R. A. Brown, "Analytical Methods for Polynuclear Aromatic Hydrocarbons in Crude Oils, Heating Oils, and Marine Tissues," Conference on the Prevention and Control of Oil Spills, Amer. Petrol. Inst., pp. 103-113 (1975).

11. P. W. Jones, R. J. Jakobsen, P. E. Strup, and A. P. Graffeo, "Chemical Characterization of Shale Oil and Related Fuels," Preprints, Div. of Fuel Chemistry, American Chemical Society, pp. 58-66 (1978).

12. E. Sawicki, T. W. Stanley, J. D. Pfaff, and W. C. Elbert, "Thin Layer Chromatographic Analysis of Polynuclear Aza- Heterocyclic Compounds," Anal. Chim. Acta 31, 359 (1964).

13. A. W. Peters and J. G. Bendoraitis, "High Resolution Mass Spectrographic Method for the Analysis of Nitrogen and Oxygen Containing Material Derived from Petroleum," Anal. Chem. 48, 966-973 (1976).

14. G. U. Dinneen, G. L. Cook, and H. B. Jensen, "Estimation of Types of Nitrogen Compounds in Shale-Oil Gas Oil," Anal. Chem. 30, 2026-2030 (1958).

CHAPTER 6

A SYSTEM FOR DATA COLLECTION AND COMPUTER PROCESSING IN OCCUPATIONAL HEALTH PROGRAMS. THE SOHIO HEALTH INFORMATION SYSTEM - PRELIMINARY REPORT

C. W. Stallard, Jr., M.D., The Standard Oil Company, Cleveland, Ohio 44115

I. PRELIMINARY DISCUSSION

BACKGROUND

The concept of profiling was gained from military experience, where mobilization of large numbers of mostly male soldiers required a comprehensible system, one that could be quickly explained to medical corpsmen, and one which provided a pattern for judgment, and a disciplined approach to the task of evaluating individuals for task assignments. The evaluation could be comprehensive, or could be relatively restricted, and any package of information could be inserted into the profile as an interim or final judgment. These concepts have served well for the past three decades of the author's experience, and coupled with code systems and data processing, constitute the present system.

DESCRIPTION AND USES OF THE SYSTEM

The medical summary sheet (Figure 1). This form encapsules virtually all of the information about an employee (or any individual, for that matter) and what is known about the status of personal health and life risks, some of which are intrinsic to the individual and are expressed in diagnostic terms or the recording of abnormal physiological values. These include historical events, family history, social behavior, and may include any variable which can be recognized and defined for the observer's judgment.

Extrinsic factors which bear on health risks are expressed in the job exposure table, but include any etiologic agent which may appear in an event recorded in the system.

MEDICAL SUMMARY

FIRST	MIDDLE	LAST NAME	SOCIAL SECURITY NO.	BIRTHDATE	EMPLY.DATE	SEX	RACE	MAR	STAT	WORK LOCATION
Harry	G.	Jones	123-45-6789	01/21/28	08/16/70	M	W	S	J	HOL80330211

STREET ADDRESS	CITY	STATE	ZIP CODE	TELEPHONE NO.
45805 Ridge Rd.	Lima	OH	44344	234-7718

EMERGENCY NOTIFICATION TO (NAME)	ADDRESS	RELATIONSHIP	TELEPHONE NO.
N. K. Jones	9240 E. Lake Blvd.	Friend	676-9231

PERSONAL PHYSICIAN (NAME)	ADDRESS	TELEPHONE NO.
Dr. Smith	2220 Brookview Rd.	676-9191

JOB TITLE	JOB CODE	JOB EXPOSURE TABLE	P	Q	R	I	N	C	E	F	O	M
Senior Engineer	89107		0	3	0	1	3	2	1	2	0	12

MEDICAL PROFILE													QUALIFICATION AND SURVEILLANCE PROGRAMS										
DATE	0	1	2	3	4	5	6	7	8	9	X	Y	DATE	P	Q	R	I	N	C	E	F	O	M
05/05/80	1	2	1	3	4	1	1	1	1	1	3	1	05/05/80	0	3	0	1	D	2	1	2	0	12

DIAGNOSTIC INDEX

DATE OF ITEM MO.	YR.	CATEGORY SOURCE	O/N	NO.	DIAGNOSIS	MP TYPE	CLASS	OCCURENCE DATE MO.	YR.	ICD-9 CM E	DIAGNOSIS	PROCEDURE
08	70	11	N	1	BURSITIS LEFT HEEL, REMOTE	1	2	00	60		72730	
11	75	61	N	2	BILATERAL ARCUS SENILIS	X	2	00	00		37141	
03	78	61	N	3	INTERNAL HEMORRHOIDS	4	2	00	77		45520	
02	79	61	N	4	CORONARY ARTERIOSCLEROSIS	4	4	00	00		41400	
02	79	61	N	5	AORTIC ATHEROSCLEROSIS	3	3	00	00		44000	
02	80	22	N	6	QUALIFICATION FOR EOP			02	80		V7030	
05	80	61	N	7	HIGH FREQUENCY HEARING LOSS, BILATERAL	X	3	05	80		38990	
05	80	61	N	8	ALLERGY TO PENICILLIN	0	2	00	00		V1400	
05	80	61	N	9	INDERAL MEDICATION					E	94200	
05	80	61	N	10	CYCLOCUMAROL MEDICATION					E	93420	
				11								
				12								
				13								
				14								
				15								
				16								
				17								
				18	FAMILY HISTORY							
03	78			19	HEART DISEASE (FATHER)						V1740	
				20								
				21								

SNOMED

NO.	TOPOGRAPHY SEM QUAL	CODE	MOD	MORPHOLOGY CODE	TUMOR	ETIOLOGY MOD	CODE	FUNCTION CODE	DISEASE CODE	PROCEDURE CODE	*	MEASUREMENT ABBREV.	ACTUAL MEASUREMENT		INITIALS
1		16560		40000											
2				50760											
3		49050		32630											
4									73540						
5									52110						
6															
7							937000000	X6030							
8						C	00087-09-2								
9							767400000			Y330					
10							841400000			Y330					
11															
12															
13															
14															
15															
16															
17															
18															
19	FH								70300						
20															
21															

MEDICATIONS, RESTRICTIONS, OTHER

	DATE MO.	YR.	SMOKE HX TYPE	LGTH.	AMT.	DRINK HX TYPE	AMT.
22	02	80	0	0	0	6	1
23							
24							
25							

COMMENTS

Figure 1

Figure 2 shows demographic data regarding an individual. This format is suitable to SOHIO's system, but is not requisite to the functioning of the system. There are:

Primary identifiers:

1. Name
2. Job title (Senior Engineer)
3. There could be an alphabetic identifier for the work location, such as the name of the plant, the town in which it was located, etc.

Secondary identifiers:

1. Social security number (or employee number, in many systems)
2. Job code
3. Coded work location

Tertiary identifiers:

1. Last known address
2. Address of friend or relative
3. Name of personal physician
4. Telephone number

The use of primary and secondary identifiers is obvious. Tertiary identifiers provide additional checks to the identity and location of the individual. Also, information of this type will allow the future investigator who needs contacts for direct follow-up to have these information sources identified on the record itself, a facility which can prove invaluable in epidemiologic studies.

JOB EXPOSURE TABLE

The headings are explained in Figure 3, and entries are explained partially by Figure 4, which deals with risk codes for specific agents.

MEDICAL SUMMARY

FIRST	MIDDLE	LAST NAME	SOCIAL SECURITY NO.	BIRTHDATE	EMPLY.DATE	SEX	RACE	MAR	STAT	WORK LOCATION
Harry	G.	Jones	123-45-6789	01/21/28	08/16/70	M	W	S	J	HOL80330211

STREET ADDRESS	CITY	STATE	ZIP CODE	TELEPHONE NO.
45805 Ridge Rd.	Lima	OH	44344	234-7718

EMERGENCY NOTIFICATION TO (NAME)	ADDRESS	RELATIONSHIP	TELEPHONE NO.
N. K. Jones	9240 E. Lake Blvd.	Friend	676-9231

PERSONAL PHYSICIAN (NAME)	ADDRESS	TELEPHONE NO.
Dr. Smith	2220 Brookview Rd.	676-9191

JOB TITLE	JOB CODE		P	Q	R	I	N	C	E	F	O	M
Senior Engineer	89107	JOB EXPOSURE TABLE	0	3	0	1	3	2	1	2	0	12

MEDICAL PROFILE												
DATE	0	1	2	3	4	5	6	7	8	9	X	Y
05/05/80	1	2	1	3	4	1	1	1	1	1	3	1

QUALIFICATION AND SURVEILLANCE PROGRAMS										
DATE	P	Q	R	I	N	C	E	F	O	M
05/05/80	0	3	0	1	D	2	1	2	0	12

DIAGNOSTIC INDEX

Figure 2

P - The physical environment
Q - Qualification programs
R - Ionizing radiation (C Subset)
I - Inhalant dusts (C Subset)
N - Noise and vibration (C Subset)
C - Chemical agents
E - Managerial examination program
F - Functional aspects of work (ERGONOMICS) (P Subset)
O - Miscellaneous programs
M - Cycle time in months

Figure 3. Job exposure codes.

0 - None, or negligible exposure.
1 - Intermittent exposure below an action level.
2 - Continuous exposure below an action level.
3 - Intermittent exposure above an action level, and below TLV or PEL.
4 - Continuous exposure above an action level, and below TLV or PEL.
5 - Intermittent exposure above TLV or PEL or occasionally above STEL.
6 - Continuous exposure above TLV or PEL or frequently above STEL.
7 - Past exposure, no current exposure.

Figure 4. Risk codes.

Each of these entries represents a composite of many factors and several inventories within each category.

P. The Physical Environment. This heading includes electromagnetic energy spectrum, thus heat, visible light, ultraviolet, infrared, also radio waves, microwaves, all non-ionizing radiant energy and ultrasound are evaluated in this heading. Climatic variables, cold, wind, atmospheric pressure, confined spaces, underground locations, driving, caisson work, high altitude, are some of the other variables of the physical environment. Risk analysis is more difficult in the physical environment because of less well-defined appraisal, thus, less quantitation. Much remains to be done to complete this entry for all jobs.

Q. Qualification. Bonafide occupation qualifications and medical standards are few. The days wherein the company's employment manager and medical advisor could evolve a manual of conditions whereby applicants could be judged for acceptance or rejection are now past, and development of applicable standards ("bona-fides") is now underway, with considerable challenge to be expected. In SOHIO's system, three qualifications are observed:

1. FAA Pilot's license renewal
2. Department of Transportation drivers
3. Emergency operation program

Two of these have directly to do with public safety. The third, which is a program of examination for management employees who would be expected to operate refineries or chemical plants in event of non-resolved labor disputes, is less obviously protective to public safety, but represents a concern for health as related to operating a closed system which could produce a major catastrophe by explosion, fire, or uncontrolled release. Concepts that lead to exclusion of employees from special job responsibilities or workplace risks which cannot be feasibly removed apply to this heading, and specific concerns such as fetotoxicity, gametotoxicity, teratogenicity will determine which individuals may undertake defined risks and responsibilities at work. Programs of this latter category are presently ill-defined.

R. Radiation. This heading defines risks due to radioactive substances, and radiant energy which is capable of producing ionization. Both physical energy and chemical characteristics are involved, so radioactive chemicals are a subset of the larger group of chemical agents, whereas x-rays are of the physical domain. Since the health effect is due to ionization of tissue, they are considered together as ionizing radiation.

I. Inhalant Dusts. These are chemical entities or agents which produce their primary health effect by depositing in the lung structure, inducing fibrosis and tissue destruction, and cause pneumoconiosis. They are thus a subset of the chemical group.

N. Noise. This subset of the physical energy spectrum is restricted to noise as related to hearing loss, and programs related to that risk and response. Ultrasound and vibration are not delineated in this grouping, but under "P".

C. Chemicals. This enormous set now contains hundreds of thousands of identifiable entities. Well defined methods for establishing risk levels are available for the environmentalist's or industrial hygienist's use. Each must be evaluated in turn, and the aggregate risk judgment appears under this heading.

E. Eligible for Managerial Examination. Most companies offer some form of health maintenance examination for key personnel, usually at decreasing intervals due to age. Including this factor in the job exposure table allows central control of the program, so that budgeting, audit of performance of costs and management of the program is made easy. Also, central review and follow-up assures that eligible employees are receiving optimum attention to their health.

F. Function. This heading contains the physical stresses of motion, and deals with kinesiologic and ergonomic variables. Also, physiological and psychological stresses, when quantifiable even by judgment, would be considered for risk codes and include under this heading. This area also is under development, as knowledge of the applicable variables is currently diffuse, and quantitation only partially achieved.

These headings (P, Q, R, I, N, C, and F) represent the judgments of analysts of the workplace, and provide a disciplined arrangement for recording the aggregate defined risks as described above.

O. This heading is left as a miscellaneous category, and is currently used for programming medical surveillance where risks have not been measured, or where a need for surveillance is discovered before risks can be adequately quantified.

M. This entry represents time in months, and triggers the next cycle of surveillance.

In addition to displaying risks in an organized form, the job exposure table codes key surveillance programs, and allows further display of the individual's status relative to job risks or evaluation programs at a specific time.

QUALIFICATIONS AND SURVEILLANCE PROGRAMS

Date. The date entered apprises the observer of the currency of the individual's status relative to job risks entered in the job exposure table. Matching numbers indicate qualification. The letter "D" indicates that the individual has a health factor which limits exposure to the agent at risk. In the example supplied, Mr. Jones is listed as "D" under "N". The reason, which appears in other areas of the summary, is hearing loss. Mr. Jones otherwise is current as of 5/05/80, and is scheduled to be re-evaluated in 12 months under the company's managerial program ("E-1"), when all data will be re-evaluated and new entries made.

MEDICAL PROFILE

The medical profile is a topographic code. The original military profile, "PULHES" indicated functional areas as in Figure 5.

P - General Physical Condition
U - Upper Extremities
L - Lower Extremities
H - Cardiovascular (and Pulmonary)
E - Eyes and Ears
S - Psychological

Figure 5. "PULHES"

This arrangement did not lend itself well to diagnostic listings related to profile entries, so the topographic code of "Standard Nomenclature of Disease and Operations" was originally used. This code structure allowed diagnoses to be weighted individually; the profile representing the most significant, or highest number, as in Figure 6.

This code is now obsolete, and its successor "SNOMED" provides different groupings and codes. Figure 7 displays this topography.

Profiling is now done, using the coded decisions as in Figure 8. Each diagnosis is weighted, and the resultant profile code number is entered under the organ system to which the diagnosis refers.

0 - Body as a whole (including the psyche (mind) and body generally)
1 - Integumentary system (including subcutaneous areolar tissue, mucous membranes of orifices, and the breast)
2 - Musculoskeletal system
3 - Respiratory system
4 - Cardiovascular system
5 - Hemic and lymphatic systems
6 - Digestive system
7 - Urogenital system
8 - Endocrine system
9 - Nervous system
X - Organs of special sense

Figure 6. Topography (SNODO)

0 - General body topography: Integumentray, hematopoietic and lymphatic systems
1 - Musculoskeletal system and soft tissues
2 - Respiratory system
3 & 4 - Cardiovascular system
5 & 6 - Digestive system
7 & 8 - Genitourinary systems and fetal structures
9 - Endocrine system
X - Nervous system and special sense
Y - Topographic regions

Figure 7. SNOMED Topography Field.

1 = Normal findings
2 = Diagnosis recorded historical importance only
No present significance
3 = Diagnosis requiring ongoing action
Chronic
Transient or undefined condition
Affects job performance
4 = Condition of disability or handicap
5 = Death
6 = T & P retirement (Non-Occ.)
7 = T & P (Occ.)
8 = Longevity retirement
9 = Termination

Figure 8. Profile Codes.

The medical summary therefore contains a great deal of information, identifying the individual, the job, the workplace, its risks, the individual's health status as current to a dated entry, and the status of the individual relative to specific requirements for qualification and to programs of medical surveillance or health maintenance.

DIAGNOSTIC INDEX

Figure 9 displays the actual diagnosis or clinical entities which have been discovered in the course of medical evaluation, and the date of any item.

Date of Item. This entry is actually the date that the item was recorded, and may be different from the "occurrence date" appearing on the same line. The occurence date locates the diagnostic entry as closely as possible in time, but usually represents a remembrance of an historical event, and is not strictly accurate as the the finite beginning of the process in question, unless it were a well defined event, such as an accident.

DATE OF ITEM MO.	DATE OF ITEM YR.	CATEGORY SOURCE	CATEGORY O/N	NO.	DIAGNOSIS	MP TYPE	MP CLASS	OCCURENCE DATE MO.	OCCURENCE DATE YR.	ICD-9 CM E	ICD-9 CM DIAGNOSIS	ICD-9 CM PROCEDURE
08	70	11	N	1	BURSITIS LEFT HEEL, REMOTE	1	2	00	60		72730	
11	75	61	N	2	BILATERAL ARCUS SENILIS	X	2	00	00		37141	
03	78	61	N	3	INTERNAL HEMORRHOIDS	4	2	00	77		45520	
02	79	61	N	4	CORONARY ARTERIOSCLEROSIS	4	4	00	00		41400	
02	79	61	N	5	AORTIC ATHEROSCLEROSIS	3	3	00	00		44000	
02	80	22	N	6	QUALIFICATION FOR EOP			02	80		V7030	
05	80	61	N	7	HIGH FREQUENCY HEARING LOSS, BILATERAL	X	3	05	80		38990	
05	80	61	N	8	ALLERGY TO PENICILLIN	0	2	00	00		V1400	
05	80	61	N	9	INDERAL MEDICATION					E	94200	
05	80	61	N	10	CYCLOCUMAROL MEDICATION					E	93420	
				11								
				12								
				13								
				14								
				15								
				16								
				17								
				18	FAMILY HISTORY							
03	78			19	HEART DISEASE (FATHER)						V1740	
				20								
				21								

MEDICATIONS, RESTRICTIONS, OTHER:

NO.	DATE MO.	DATE YR.	SMOKE HX TYPE	SMOKE HX LGTH.	SMOKE HX AMT.	DRINK HX TYPE	DRINK HX AMT.
22	02	80	0	0	0	6	1
23							
24							
25							

CLV-8767-1-A

Figure 9

Category, Source O/N. It is quite necessary to define diagnostic entries relative to the period of employment. Those existing prior to employment are identified by using source codes: in this instance code "11" indicates that this historical entry was made at the time of the employment examination (08/70) but had occurred ten years before (00/60) and was not now present or was of historical interest only.

Subsequent entries are made as conditions are discovered in subsequent examinations (Code 61 for managerial examinations). None of the conditions were related to occupation (N under O/N).

Profiling: MP/Type/Class. Each diagnostic entry is weighed according to the value system in Figure 8. In many state worker's compensation statutes, the employer may list those diagnoses which are regarded as partially disabling. (Class or Profile "4"). These conditions are not the employer's liability, but factored into a disability evaluation, will allow payments from second injury funds. The profile is thus a source for compilation of lists of employees who are or should be so listed. Additionally, conditions classified may be computed rapidly, together with displays relative to job assignment, race, sex, age, and specific diagnosis as factors pertaining to job assignment.

Obvious entries such as drugs taken, smoking history, and alcohol usage help to define the "internal environment" of the individual, and are of value in approaching the complex issue of causation of disease.

Although the assistant secretary for OSHA has indicated that occupational and nonoccupational disease are virtually indistinguishable, maintenance of records of this completeness clearly define the relationship, or absence thereof, of job exposure to alterations in health.

CODE STRUCTURES

SOHIO's health information was designed to meet several criteria, as follows:

1. It must lend itself to clinical use, and be comprehensive to a company's medical staff.

2. It must be problem-oriented, and arranged in topological format so that data elements are immediately visible, and that the timeliness, or current status of the record, is readily apparent.

3. It must incorporate the concept of profiling.

4. The clinical record should display a representation of cause and effect, in topological form.

5. It must use commonly available code structures wherever possible, and any unique code structures must be capable of easy substitution by other users.

   A. Codes unique to the SOHIO system include:
      1) The company's location codes, which are derived for payroll and accounting purposes
      2) Employee numbers
      3) Job code numbers
      4) Source code numbers - used for identifying the origin of diagnostic information
      5) Medical profile codes
      6) Environment risks

   B. Codes that are not unique include the ICD-A 9 code for morbidity and mortality, and the larger and more comprehensive SNOMED code (Figure 10).

Since SNOMED is intended primarily for the clinician's use, it required additional qualifying entries to fit it to SOHIO's system. These appear adjacent to the major headings of SNOMED, and consist of the following:

1. Semantic Qualifier:
   Allows the coder to incorporate graduation of meaning and nuances, and linkages indicating the relationship between one diagnosis and another.

eg. 1) CD=Clinical diagnosis
    2) FH=Family history of
    3) HR=High risk of
    4) LD=Laboratory diagnosis
    5) PH=Past history of
    6) PT=Primary tumor in

eg. FH/T28100/M8140/3/P

Person had primary Adenocarcinoma of the left lung (Family HX of lung CA is noted)

2. Topography Code:

   Site of the body where problem occurs.

eg. T-11410 = Humerus, NOS (Not otherwise specified)

3. Topographical Modifier:

   A. Many specified topography codes are present for paired (eg. kidneys) and multiple (eg. ribs)

NO.	TOPOGRAPHY SEM QUAL	TOPOGRAPHY CODE	TOPOGRAPHY MOD	MORPHOLOGY CODE	TUMOR	ETIOLOGY MOD	ETIOLOGY CODE	FUNCTION CODE	DISEASE CODE	PRO-CEDURE CODE	*	MEASURE-MENT ABBREV.	ACTUAL MEASURE-MENT		INITIALS
1		16560		40000											
2				50760											
3		49050		32630											
4									73540						
5									52110						
6															
7							9370000000	X6030							
8							C00087-09-2								
9							7674000000			Y330					
10							8414000000			Y330					
11															
12															
13															
14															
15															
16															
17															
18															
19	FH								70300						
20															
21															

COMMENTS

Figure 10

structures. However, codes are not present for designating side of body such as right humerus or right tenth rib; therefore, an additional modifier field can be used. (Position #1)

eg. POSITION #1

0=Null
1=Right
2=Left
3=Bilateral

B. Some codes lack information such as the information in Position #2.

eg. POSITION #2

0=Null	6=Upper/superior
1=Proximal	7=Lower/inferior
2=Middle	8=Anterior/ventral
3=Distal	9=Posterior/dorsal
4=Medial	U=Segment
5=Lateral	

eg. T-10355/<u>11</u> = 5th rib, <u>right proximal</u>

4. <u>Morphology</u>:

Describes or names the pathological changes or processes occurring in cells, tissues, or organs.

eg. M-40000 = inflammation

5. <u>Tumor Verification</u>:

The fifth digit of a morphology code for cancer (M8000 through M9990) specifies whether the tumor is benign (M8000/6). To save space, the SNOMED book lists the fifth digit of all cancers as 3. If the coder wishes to record a metastatic tumor, the fifth digit must be changed from 3 to 6. Therefore, the verification letter is a single alphabetic letter (eg. B=Benign, P=Primary, X=metastatic, etc.) immediately following the fifth digit and is used to verify that the fifth digit has not been inadvertently recorded as 3.

eg. T28100 - M8140/3P Primary site adenocarcinoma of the left lung.

T28100 - M8140/6X Metastatic adenocarcinoma, left lung.

6. <u>Etiology Modifier</u>:

The etiology modifiers may be used to identify drugs used in therapy or reagents used in a laboratory test rather than an etiology of disease. This field also designates if the number in the etiology is a CAS number rather than a SNOMED code by placing a "C" in the field.

eg. 0=Null
1=Etiologic Agent
2=Therapy for condition
3=Reagent for test
4=Whatever else is decided
C=CAS number

*NOTE: Could be changed to alpha characters.
eg. /1/E-4200=Penicillin cause of a problem such as an allergic reaction and not treatment for a condition.

7. <u>Etiology Code With Incorporation of CAS Number</u>:

Lists names of etiologic agents or causes of disease or morphologic change. Since many etiologic agents can be chemicals, the CAS numbers have been incorporated into this field to replace SNOMED chemical codes, because SNOMED chemical listing is incomplete. It would be more economical to incorporate the CAS numbers into the etiology rather than to create another additional 10-space field for the CAS number.

eg. E-9440 = firearm - use SNOMED because an object was the etiologic agent

/C/E-000107131 = Acrylonitrile - use CAS number because chemical was the etiologic agent.

8. <u>Function Code</u>:

(Remains the same with no new additional fields.) Describe functional disorders in the pathophysiological sense.

9. <u>Disease Code</u>:

(Remains the same with no new additional fields.) The classification component of the system.

eg. D-8251 = Classic migraine

10. Procedure Code:

Administrative, Diagnostic, Therapeutic, and preventive procedures done.

eg. P-2885 = RBC count

11. Procedure Modifier:

These modifiers allow specification of the indication for the performance of a procedure (diagnostic, therapeutic).

eg. 1=Therapeutic
2=Diagnostic
3=Follow-up
4=Screening
5=Investigative
6=Administrative
7=Medico-Legal
8=Educational
9=Public Health

eg. T-56970PX750/4 "Liver imaging with vascular flow" test is done on employees exposed to certain chemicals. Coder wants to indicate it was a screening program.

12. Asterisk Modifier*:

An asterisk is filled in by the coder if the date from the procedure performed is going to be recorded. It will probably only be necessary to record information if the results were abnormal.

eg. P-4020/*/MG0364 High cholesterol level - decided that it should be recorded as found in test results.

13. Abbreviation Denoting the Unit of Measurement Used in the Procedure:

Any unit of measurement can be abbreviated and put in this field in order to identify procedure results in units and numbers.

eg. P-4020/*/MG/0364 = Cholesterol, 364 mg

14. Raw Data/Actual Measurement Taken:

The actual numerical result obtained from a procedure performed.

eg. P-4154/HPF/0001 Urinalysis
Red blood cells = 1 HPF

Additionally, modification of SNOMED for code definitions of environmental exposure may require revision of the codes themselves. To date, a mixture of SNOMED codes, ANSI Z-16 codes for industrial injury, and chemical abstract registry numbers for chemicals, have caused us to expand the etiology section of SNOMED to accommodate these other codes, which are either larger code systems, more explicitly related to need, as in industrial accident reporting, or are preferred over SNOMED's code selections. Since all of these entries appear under the single field, Etiology, and the CAS registry numbers occupy a position of prominence, the system has worked well, although additional effort is needed to perfect it.

SOHIO's HEALTH INFORMATION SYSTEM CONSISTS OF SIX BASIC FILES:

File 1 - Master file of employee data extracted from payroll and personnel systems, and supplying all demographic identifiers. This file also contains the medical profile and job exposure table.

File 2 - Contains all diagnostic entries coded to ICDA-Rev. 9 and to the modified SNOMED code system. The medical profile is repeated as type/class entries.

File 3 - A job exposure inventory. This file displays risk assessment and code numbers for each physical and chemical agent as applicable to each individual job, and is incomplete for many locations at this time. This file accesses file 4.

File 4 - Master code inventory, which lists all chemicals present in the company, identifying mixtures by a four digit master code, and generic chemicals by CAS number. The SOHIO master code is incorporated into the CAS number field when identification is incomplete. See Figures 11 and 12. A chemical inventory by job code resembles Figure 13 and may be related to job codes by referring to lists such as Figure 14, which provides name and CAS number for ready input.

File 5 - Industrial hygiene data file. This file is constructed from actual measurements, and is related to area sampling or to individual employees. Ultimately, sampling strategy will allow that measured exposures are all related to individual employees as a function of job exposure. See Figures 15 and 16.

File 6 - Toxicological information file. This file, under development, applies the concept of topographic profiling to animal toxicology, where organ specificity or function

03/24/80

VISTRON MASTER CODE REPORT: ALPHA BY ITEM NAME

B C	MASTER CODE	EB/ IR	U S	ITEM NAME / ALTERNATIVE DESIGNATION			PRODUCT FILE NO	CAS NUMBER	STRT DATE	STOP DATE	OTHR CODE	PROD. CODE	A	PK CD
V	7010	AAM		ACETIC ACID			250.020	000064-19-7	-	-				
V	7010	AGM		ACETIC ACID GLACIAL			250.020	000064-19-7	-	-	1005			
V	8015	AAM		ACETONE			260.000	000067-64-1	-	-				
V	6303	AAM		ACETONITRILE			230.040	000075-05-8	-	-				
V	6303	ABM		ACETONITRILE			230.040	000075-05-8	-	-		72000	0	11
				ACETONITRILE	BLK									
V	6303	ACM		ACETONITRILE			230.040	000075-05-8	-	-		72100	0	11
				ACETONITRILE	DRM									
V	6303	AGM		ACETONITRILE AQUEOUS			230.040	000075-05-8	-	-		72050	0	11
				ACETONITRILE AQUEOUS										
V	6303	AZX		ACETONITRILE COMMODITY CODE		28-181-73	230.040	000075-05-8	-	-		72099	0	11
				ACETONITRILE 28-181-73										
V	6303	AKM		ACETONITRILE CRUDE			230.040	000075-05-8	-	-		72200	0	11
				ACETONITRILE CRUDE										
V	8020	AAM		ACETYLENE			260.000	000074-86-2	-	-				
V	7020	AAM		ACROLEIN			250.030	000107-02-8	-	-				
V	6308	AAM		ACRYLAMIDE			230.040	000079-06-1	-	-				
V	6308	ABM	D	ACRYLAMIDE			230.040	000079-06-1	-	-		77100	0	11
				ACRYLAMIDE	BLK									
V	6308	AEM		ACRYLAMIDE BRICKS			230.040	000079-06-1	-	-		77000	0	11
				ACRYLAMIDE BRICKS	BAG									
V	6308	AGM		ACRYLAMIDE CRYSTAL			230.040	000079-06-1	-	-		77050	0	11
				ACRYLAMIDE CRYSTAL	BAG									
V	6308	AZX		ACRYLAMIDE SOLUTION COMMODITY CODE		28-999-91	230.040	000079-06-1	-	-		77099	0	11
				ACRYLAMIDE SOLN. 28-999-91										
V	6308	ASM		ACRYLAMIDE SOLUTION 50%			230.040	000079-06-1	-	-		77110	0	11
				ACRYLAMIDE SOLN. 50%	BLK									
V	6308	AUM		ACRYLAMIDE SOLUTION 50%			230.040	000079-06-1	-	-		77120	0	11
				ACRYLAMIDE SOLN. 50%	DRM									
V	7025	AAM		ACRYLIC ACID			250.050	000079-10-7	-	-				
V	6312	AAM		ACRYLONITRILE			230.040	000107-13-1	-	-	1000			
V	6312	ABM		ACRYLONITRILE			230.040	000107-13-1	-	-		71000	0	11
				ACRYLONITRILE	BLK									
V	6312	ACM		ACRYLONITRILE			230.040	000107-13-1	-	-		71100	0	11
				ACRYLONITRILE	DRM									

Figure 11

03/24/80 VISTRON MASTER CODE REPORT: NUMERIC BY MASTER CODE PAGE 65

B C	MASTERE CODE	B/ I	U R S	ITEM NAME ALTERNATIVE DESIGNATION		PRODUCT FILE NO	CAS NUMBER	STRT DATE	STOP DATE	OTHR CODE	PROD. CODE	A	PK CD	ANAL N	ANAL P	ANAL K
V	7245	A	AM	DIBUTYLTHIOSULFATE		250.020	X17245-00-0	-	-					.	.	.
V	7260	A	AM	DILAURYLDITHIOPROPIONATE		250.050	000123-28-4	-	-					.	.	.
V	7260	A	NM	NAUGARD DLTDP	UNIROYAL	250.050	000123-28-4	-	-	8042				.	.	.
V	7280	A	AM	DOW FG 10 ANTIFOAM ANTIFOAM FG 10 DOW	DOW CHEM.	250.050	X07280-00-0	-	-	8026				.	.	.
V	7298	A	AM	ETHANOL AMINE MONO MONOETHANOL AMINE		250.030	000141-43-5	-	-	1609				.	.	.
V	7310	A	AM	FORMALDEHYDE		250.030	000050-00-0	02-80	-					.	.	.
V	7310	A	FM	FORMALDEHYDE SOLUTION		250.030	000050-00-0	02-80	-	1279				.	.	.
V	7318	A	AM	GAS NATURAL METHANE		250.030	008006-14-2	-	-	0010				.	.	.
V	7345	A	AM	HYDROCHLORIC ACID HYDROGEN CHLORIDE		250.020	007647-01-0	-	-	1004				.	.	.
V	7350	A	AM	HYDROGEN		250.030	001333-74-0	-	-					.	.	.
V	7358	A	AM	HYDROXYLAMINE DIETHYL DIETHYLHYDROXYLAMINE		250.050	003710-84-7	-	-					.	.	.
V	7358	A	PM	PENNSTOP RC 1866	PENNWALT CORP	250.050	003710-84-7	-	-	8034				.	.	.
V	7385	A	AM	IRON NITRATE FERRIC NITRATE		250.020	010421-48-4	-	-	0300				.	.	.
V	7390	A	AM	IRON SULFATE FERROUS SULFATE		250.020	007720-78-7	-	-					.	.	.
V	7390	A	EM	IRON SULFATE 20 LB FERROUS SULFATE 20 LB		250.020	007720-78-7	-	-	1276				.	.	.
V	7390	A	GM	IRON SULFATE 50 LB FERROUS SULFATE 50 LB		250.020	007720-78-7	-	-	1277				.	.	.
V	7395	A	AM	ISOBUTYLENE		250.030	000115-11-7	-	-	0500				.	.	.
V	7450	A	AM	MANGANESE NITRATE		250.020	010377-66-9	-	-					.	.	.
V	7465	A	AM	MELAMINE		290.000	000108-78-1	-	-					.	.	.
V	7465	A	MM	MELAMINE TRANSFER TO PRO BRUSH MELAMINE PRO BRUSH		290.000	000108-78-1	-	-		78500	1	11	00.00	00.00	00.00
V	7475	A	AM	MERCAPTAN T-DODECYL T-DODECYL MERCAPTAN		250.050	025103-58-6	-	-	8021				.	.	.
V	7490	A	AM	METHYL ACRYLATE		250.050	000096-33-3	-	-	8011				.	.	.
V	7500	A	AM	MOLYBDENUM OXIDE MOLYBDIC OXIDE		250.020	001313-27-5	-	-	1608				.	.	.
V	7505	A	AM	MONAWET MO 70 R DOSS	MONA INDUSTRIES	250.050	X07505-00-0	-	-	8032				.	.	.
V	7505	D	AM	DIOCTYL SODIUM SULFOSUCCINATE	COMPONENT	250.050	001639-66-3	-	-					.	.	.
V	7505	P	AM	PROPYLENE GLYCOL	COMPONENT	250.050	000057-55-6	-	-					.	.	.
V	7515	A	AM	NAUGARD TKB	UNIROYAL	250.050	X07515-00-0	-	-	1631				.	.	.

Figure 12

Resin Mixer

CO.	PG.	DEPT.	DIV.	UNIT	JOB CODE	CAS #	MEASUREMENT DATE MO.	DAY	YEAR	ACT. TEST	RISK	COMMENTS
HO	C	08	15	1036	03190	000057-55-6	05	20	80	E	1	
						000085-44-9				E	1	
						000108-31-6				E	1	
						000100-42-5				E	3	SKIN
						000080-62-6				E	1	
						000123-31-9				E	0	
						001838-02-9				E	0	
						000112-00-5				E	0	
						000557-09-5				E	0	
						002440-22-4				E	0	
						000126-30-7				E	1	
						000095-71-6				E	1	
						025265-71-8				E	1	
						000682-09-7				E	1	
						007631-86-9				E	0	SKIN
						000067-64-1				E	4	SKIN
						000094-36-0				E	4	SKIN
						000080-15-9				E	3	SKIN
						001338-23-4				E	3	SKIN
						065997-17-3				E	1	

Figure 13. Chemical Inventory by Job Code - Filon, Covington.

COMPONENT	CAS REGISTRY NO.	COMPONENT	CAS REGISTRY NO.
Acetone	67-64-1	Methyl Ethyl Ketone	108-10-1
Adipic Acid	124-04-9	Methyl Ethyl Ketone Peroxide	78-93-3
Aerosil (Silica, Amorphous fumed)	7631-86-9	Methyl Isobutyl Ketone	1338-23-4
Azelaic Acid	123-99-9	Methyl Methacrylate	108-10-1
Benzoyl Peroxide	94-36-0	Neopentyl Glycol	80-62-6
n-Butyl Acrylate	141-32-2	Pentaerythritol	126-30-7
Cellosolve Acetate	111-15-9	Phthalic Anhydride	115-77-5
Copper Naphtenate	1838-02-9	Propylene Glycol	85-44-9
Cumene Hydroperoxide	80-15-9	Styrene	57-55-6
Cyclohexane Dimethanol	105-08-8	Tertiary - Butyl Perbenzoate	100-42-5
Cyclohexanone	108-94-1	Tetrabromophthalic Anhydride	614-45-9
Diallyl Phthalate	131-17-9	Tetrachlorophthalic Anhydride	632-79-1
Dibromoneopentyl Glycol	3296-90-0	Tinuvin P (A-50)	117-08-8
Dicyclohexyl Phthalate	84-61-7	Tolu HQ	2440-22-4
Diethylene Glycol	77-73-6	Trimethyldodecylammonium Chloride	95-71-6
Dimethyl Methyl Phosphonate	111-46-6	Triethyl Phosphate	112-00-5
Dipropylene Glycol	756-79-6	Trimethylolpropane	78-40-0
Ethylene Glycol	25265-71-8	Vinyl Toluene	77-99-6
Fibrous Glass	107-21-1	Zinc Octoate	682-09-7
Fumeric Acid	65997-17-3		25013-15-4
Glycerine	110-17-8		558-09-5
Hydroquinone	56-81-5		
Isophthalic Acid (PARA)	123-31-9		
Maleic Acid	121-91-5		
Maleic Anhydride	110-16-7		

Figure 14. Alphabetical listing of all chemicals utilized at Filon, Covington identity and CAS registry numbers of components of resins and chemicals used or stored at Filon, Covington.

INDUSTRIAL HYGIENE SURVEY

FACILITY			ATMOSPHERIC CONDITIONS			NO
Filon Plant, Covington, Ky.	NAME	DATE	TEMPERATURE	HUMIDITY	BAROMETRIC PRESSURE	
SAMPLED	N. K. Visnapuu	11/14/79	75 F (Line area) 64 F (Mat rm)	36% (Line area) 28% (Mat rm.)	29.4	
ANALYZED			WEATHER DESCRIPTION:			
REPORTED			WIND: SPEED		DIRECTION	

COLLECTION MEDIA SAMPLING EQUIPMENT AND FLOW RATE (L/MIN.)	START, STOP ELAPSED TIME	SAMPLE VOLUME (LITERS)	SAMPLE LOCATION OPERATING CONDITIONS	EMPLOYEE NAME EMPLOYEE NUMBER SOCIAL SECURITY NO. JOB TITLE AND/OR CODE	SAMPLE NUMBER	SAMPLE RESULTS ppm ___ IN AMBIENT AIR
BDX-44 2-77-84 1.6 l/min. 0.8μ PVC	7:51 a.m. 10:38 a.m. 167 min.	48428 56367 7939 267.2 l	Mat room operator	Karen Vance 45396 402-64-6517 Line Person Code: 02460/0000	1114799 0.01073 0.01079 0.00006 g	Total dust 0.22 mg/cu
BDX-44 2-77-84 1.6 l/min. 0.8μ PVC	12:49 p.m. 3:57 p.m. 188 min.	56367 62203 5836 300 l	Mat room operator	Karen Vance 45396 402-64-6517 Line Person Code: 02460/0000	1114798 0.00980 0.01035 0.00055 g	Total dust 1.83 mg/cu
BDX-44 1-77-491 1.6 l/min. 0.8μ PVC	8:04 a.m. 2:55 p.m. 657.6 l	77441 96759 19318 657.6 l	Underneath the mat rm. (filter room)--on the floor.	Area sample--mat room operator cleans this room once per shift.	1114797 0.00958 0.00997 0.00039 g	Total dust 0.59 mg/cu
BDX-44 2-77-114 1.6 l/min.	7:54 a.m. 2:23 p.m. 389 min.	90956 05807 14851 622.4 l	On the corner desk-5 ft. height. Matched weight AA.8μ filters	Area sample--mat room.	11147910 0.04894 0.04850 0.00044 g	Total dust 0.71 mg/cum
BDX-44 1-77-334 1.6 l/min.	8:11 a.m. 2:23 p.m. 372 min.	66997 81703 14706 595.2 l	On the corner desk--5 ft. height. Matched weight AA.8μ filters.	Area sample--mat room. Open face sample	11147911 0.04853 0.04844 0.00009 g	Total dust 0.15 mg/cum

CLV-8876

DATE Nov. 14, 1979 SIGNED Naima Visnapuu, Industrial Hygien[ist]

Figure 15

INDIVIDUAL EXPOSURE RECORD

NO.	11/14/79
PLANT LOCATION: COVINGTON, KY	CODE: HO-C-80-15
OPERATING UNIT: Filon	CODE: 1036

DATE OF MEASUREMENT: 11/14/79	AGENT MEASURED: FIBROUS GLASS DUST

EMPLOYEE: K. G. VANCE	EMPLOYEE NO.: 45396

SOCIAL SECURITY NO: 402-64-6517	JOB TITLE: LINE PERSON (MAT ROOM OPERATOR)	CODE: 02460/0000

SAMPLING ANALYTIC METHOD

BDX-44 SAMPLING PUMP, 08 ,µ filter, gravimetric analysis

INSTRUMENTATION

LOWEST LIMIT OF DETECTION

0.01 mg

CALIBRATION DATE

11/14/79

ATMOSPHERIC CONDITIONS:	WIND	WEATHER	TEMPERATURE	HUMIDITY	BAROMETRIC PRESSURE
			75F (Line Area) 64F (Mat Room)	36% 28%	29.4

	TWA	CEILING
SAMPLE TYPE:		
SAMPLE DURATION:	TIME ON 1) 7:51 a.m. 2) 12:49 p.m.	TIME OFF 1) 10:38 a.m. 2) 3:57 p.m.
ELAPSED TIME	FLOW RATE 1) 1.6 l/min. 2) 1.6 l/min.	VOLUME (LITERS) 1) 267.2 l 2) 300.0 l

RESULTS

1) TOTAL DUST 0.22 mg/cum.

2) TOTAL DUST 1.83 mg/cum.

EXPOSURE OF EMPLOYEE

An average of three area samples, 0.48 mg/cum, used for 125 minutes

$$\frac{(0.22 \times 167) + (1.83 \times 188) + (0.48 \times 125)}{480} = \frac{0.92 \text{ mg/cum}}{8 \text{ hr. TWA}}$$

TWA *1979) = 10 mg/cum/8-hr TWA

RISK CODE IS I- 1 For Job Exposure table, this job

SAMPLED BY: N. K. Visnapuu, Industrial Hygienist	DATE: 11/14/79
ANALYST	AUTHENTICATED: Naima Visnapuu, Industrial Hygienist, DATE: 11-15-79

CLV-8875

Figure 16

can be identified. It will display the intrinsic risk of individual chemicals primarily, with emphasis on neurotoxicity, myelotoxicity, hepato- and nephrotoxicity, but also to display genetic effects leading to mutagenicity, carcinogenicity and reproductive toxicity. This file is expected to generate computer reports to supervision and to employees, customers, or the public where the company's products are involved.

Extensive discussion of the industrial hygiene and toxicological functions of the system is not within the purview of this report, but will appear in later submissions.

CHAPTER 7

CHRONIC DERMAL TOXICITY OF PARAHO SHALE OIL AND DISTILLATES*

J. M. Holland, L. C. Gipson, M. J. Whitaker, T. J. Stephens, and G. M. Clemmer, Biology Division, Oak Ridge National Laboratory, Oak Ridge, Tennessee 37830

D. A. Wolf, Computer Sciences Division, Union Carbide Corporation, Oak Ridge, Tennessee 37830

INTRODUCTION

The purpose of this experiment was to estimate the relative skin oncogenicity of a Paraho process shale oil and related distillates. A secondary objective was to determine whether materials applied to the skin were absorbed, ingested or possibly inhaled in quantities sufficient to induce systemic toxicity.

The results obtained reflect continuous exposure of mice for more than 12 months. Exposures will continue for another 12 months and for this reason the current results should be treated as preliminary pending terminal sacrifice and examination of surviving animals.

MATERIALS AND METHODS

Animals

Equal numbers of male and female C3Hf/Bd inbred mice were randomly distributed to treatment groups at 10 weeks of age. Mice were housed 5 per cage on sterilized hardwood chip bedding and given free access to sterilized Purina 5010C diet and hyperchlorinated-acidified water.

*Research supported by the Office of Health and Environmental Research, U. S. Department of Energy, under contract W-7405-eng-26 with the Union Carbide Corporation.

Treatment groups, concentrations and the number of animals in each group are given in Table I. To determine whether the interval between successive doses (dose rate) made any difference in toxicity the materials were applied on both a three times and a two times weekly schedule. For materials capable of evoking skin tumors on the 3 times weekly schedule we plan to stop treatment while continuing twice weekly exposures until the same number of dosages have been applied on both schedules. By contrasting the cummulative hazard rates as a function of dose rate, material which show a lesser degree of dose rate dependence can be identified. It is these mixtures which represent the greatest potential occupational hazard because even infrequent or widely spaced exposure could represent a finite risk due to either to long skin residence times or irreversible tissue injury with each exposure.

Application of Test Materials

An appropriate amount of each test material was weighed into a number of glass vials. The vials, containing a known weight of material, were kept at 4°C in an explosion proof refrigerator. Each week a vial was removed and a volume of spectrograde cyclohexane added to yield the highest concentration required for the study. Lower concentrations of the same material were made from this sample. Benzo(a)pyrene solutions were also made fresh weekly.

Hair was removed from the upper back with electric clippers. Materials were applied using a 50 μL micropipette. Concentration of each test material was determined as the maximum permitted either by physical characteristics or acute local or systemic toxicity. Whenever possible the materials were applied without dilution in order to demonstrate toxicity without possible influences by the solvent.

Parameters of Acute and Chronic Dermal Toxicity

The area directly exposed to the test materials is monitored for non-specific inflammatory, hyperplastic and neoplastic changes and these lesions are recorded for each animal at the time of treatment or when the mouse is reshaved. In addition to local effects on skin texture, thickness, pigment and hair growth, there are potential systemic toxic effects due to absorption of components through the skin, ingestion or a combination of both. Also, for the more volatile components of the mixture, there is a distinct possibility that inhalation exposure occurs. Therefore a dermal "carcinogenesis" bioassay results in a multipath, natural route, exposure of the animal to the test material.

Table I. Experimental Design

Material (Repository No.)	Concentrations (w/v%)		No. of Mice/Dose		Total No. of Mice at Risk
	3/wk	2/wk	3/wk	2/wk	
Crude shale oil (4601)	50	50,25,12	30	20	90
Hydrotreated shale oil (4602)	50	50,25,12	30	20	90
Hydrotreated residue (4607)	50	50,25,12	30	20	90
Shale oil JP5 (4608)	100,50	100,50,25,12	30	20	140
Shale oil JP8 (4609)	100,50	100,50,25,12	30	20	140
Shale oil DFM (4610)	100,50	100,50,25,12	30	20	140
Petroleum JP5 (4614)	100,50	100,50,25,12	30	20	140
Petroleum JP8 (4615)	100,50	100,50,25,12	30	20	140
Petroleum DFM (4616)	100,50	100,50,25,12	30	20	140
Benzo(a)pyrene	0.1,0.05,0.025	-	30	-	90
Cyclohexane	100	-	30	-	60

Given the protracted nature of the exposure there is ample time for cummulative toxicity to become manifest. This toxicity can be reflected as decreased motor activity, neurological signs, or more readily quantifiable effects on body weight, organ weight or in the extreme, an increased mortality rate. An additional physiologic parameter is the average daily water intake required to maintain a favorable water balance. And finally tissues in animals that die or that survive to the end of the study are examined systematically and any lesions or abnormalities observed are taken for histologic examination.

Water intake for individual cage groups was determined by placing a known quantity of drinking water into the standard water bottle and measuring the remainder at the end of one week. Bottles were placed on the cages and removed at the same time for all treatment groups and vehicle controls. No attempt was made to account for spillage because it was assumed that this would be comparable between treated and vehicle control groups. Twenty-four-hour water intake, urine output and feces weight was also determined for individual mice from representative treated and vehicle control groups by placing them into metabolism cages.

Statistical Analysis of Skin Tumor Responses

The temporal distributions of skin tumor latencies, for all groups in which two or more animals had developed grossly typical skin tumors at the site of application, were fit using both the non-parametric Kaplan-Meier procedure and the parametric Weibull model relating dose in mg per week to the expectation of survival without a treatment related skin tumor. Interim relative potency estimates were obtained by comparing the dose response obtained with benzo(a)pyrene to that observed for the unknown through the current point in the experiment. Details concerning the approach used and other examples can be found elsewhere.[1-4] As additional animals develop tumors in the various treatment groups we do not anticipate that the estimated relative potencies will change substantially, however, the confidence limits will narrow.

For continuous variables, such as body weight and water consumption, group means were compared using student's t-statistic with significance at the 0.05 level.

RESULTS AND DISCUSSION

Skin Tumor Induction

Both neoplastic and non-neoplastic changes have been induced to different degrees by the various treatments.

Grossly typical neoplasms have been observed in animals exposed to the Shale Oil Crude, the Hydrotreated Crude and the Hydrotreated Residue on both the three times and the two two times weekly dosage schedule (Table II). At the two highest benzo(a)pyrene dose levels all the animals at risk have developed a skin neoplasm and thus exposure was discontinued. Neither tumors or non-specific inflammatory lesions have been observed in mice exposed to the cyclohexane vehicle.

Interim calculation of relative skin carcinogenicities was accomplished by comparison of the dose responses obtained for the benzo(a)pyrene reference carcinogen to that obtained for each of the unknowns, with dose rate expressed on a mg per week basis. The observed proportion of mice surviving without a skin tumor in cohorts exposed to tumorigenic materials is shown in Figures 1-4 for the benzo(a)pyrene, Shale Oil Crude, Hydrotreated Shale Oil and Hydrotreated Residue, respectively. The solid, step curves represent the Kaplan-Meier estimates and the superimposed dashed lines the fitted Weibull with the shape and location parameters of this model constant for all groups. Wherever the solid line stops represents the cut-off for the interim relative potency estimates for that dose-treatment group.

By comparison of the fitted Weibull curves, relative potencies are obtained for each of the positive unknowns in terms of benzo(a)pyrene dose equivalents. Table III lists these relative potencies together with upper 95% confidence limits. These values can be interpreted as the relative carcinogenic efficiency of the test material to an equivalent amount of BaP when both are applied continuously. In order of decreasing potency are the Shale Oil Crude (1/998), the Hydrotreated Residue (1/1622) and the Hydrotreated Shale Oil (1/2780) applied on the same 3 times weekly schedule as were the BP positive controls. Since there were no BP twice weekly exposure groups, the twice weekly and three times weekly relative potency estimates are not entirely comparable. However, the potency estimates are based upon dosage per week and thus difference in the schedule of application would not affect the calculation. It is noteworthy that reducing the dose rate by a factor of 1/3 reduced the tumorigenicity of the Shale Oil Crude by a factor of 1.7, the Hydrotreated Residue 2.8, and the Hydrotreated Shale Oil, 3.2. The implication of this is that the probability of tumor occurrence decreases multiplicatively as the dose rate decreases arithmetically.

Table II. Skin Tumor Frequency in Mice Exposed Either 3 Times or 2 Times Weekly to Graded Doses of Shale Oil Crude, Hydrotreated Shale Oil or Hydrotreated Residue for 10 Months

Material	Concentration (w/v%)	Percent with Tumor 3/wk	Percent with Tumor 2/wk	Average Time to 1st Tumor (days±S.E.) 3/wk	Average Time to 1st Tumor (days±S.E.) 2/wk
Shale oil crude	50	93	75	206 (6)	261 (11)
	25	nt[a]	35	-	271 (18)
	12.5	nt	10	-	278 (52)
Hydrotreated shale oil	50	46	5	228 (12)	280
	25	nt	0	-	-
	12.5	nt	0	-	-
Hydrotreated residue	50	70	20	192 (11)	270 (26)
	25	nt	0	-	-
	12.5	nt	0	-	-
Benzo(a)pyrene	0.1	100	nt	169(6)	-
	0.05	100	nt	215 (4)	-
	0.025	67	nt	254 (4)	-

[a]Not tested.

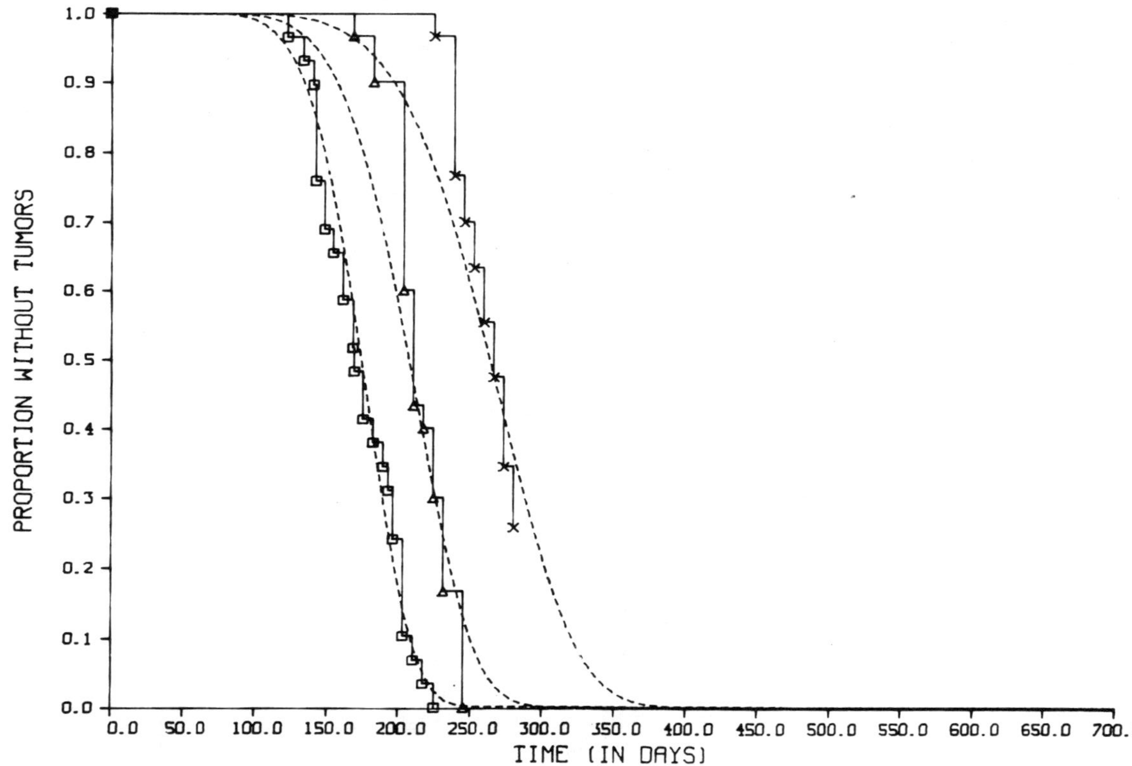

Figure 1. Proportion of mice surviving without a skin tumor following three times weekly application of benzo(a)pyrene at 0.15 (□), 0.075 (△) and 0.035 (X) mg per week. Solid lines refer to the Kaplan-Meier estimates and the dashed lines the fitted Weibull.

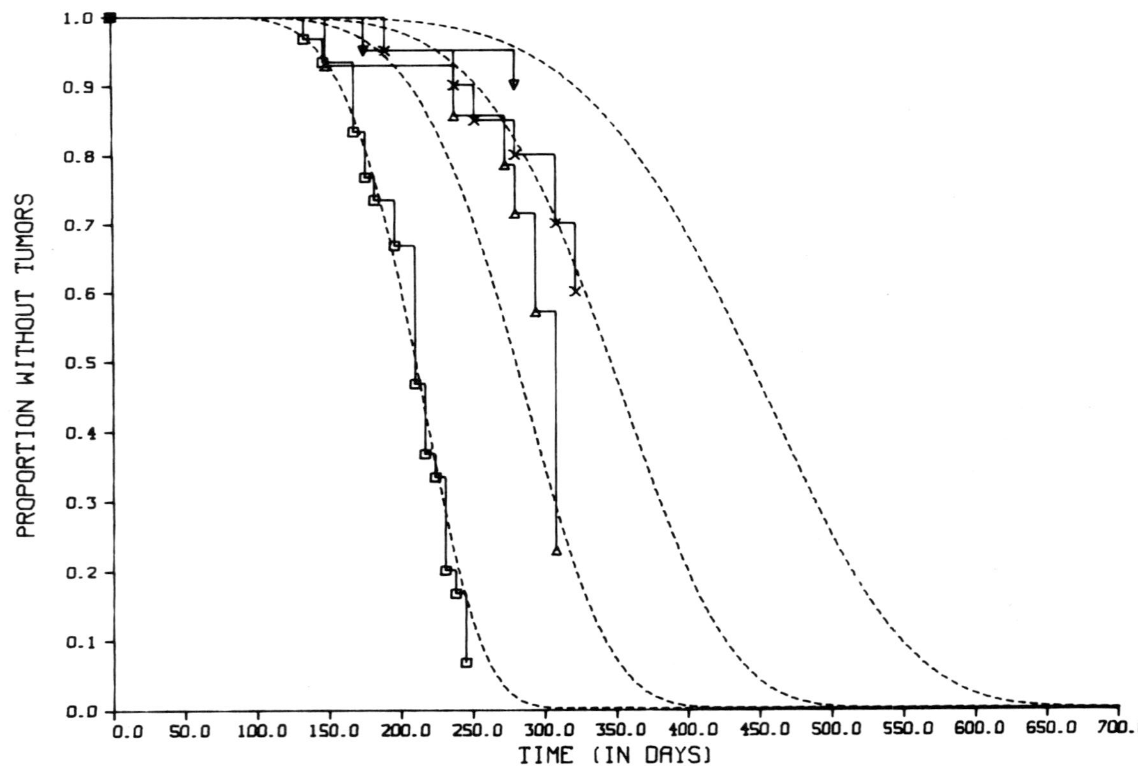

Figure 2. Proportions of mice surviving without skin tumor following application of shale oil crude on a three times weekly schedule at 75 (□) mg per week and 50 (Δ), 25 (X) and 12 (▽) mg per week, twice weekly.

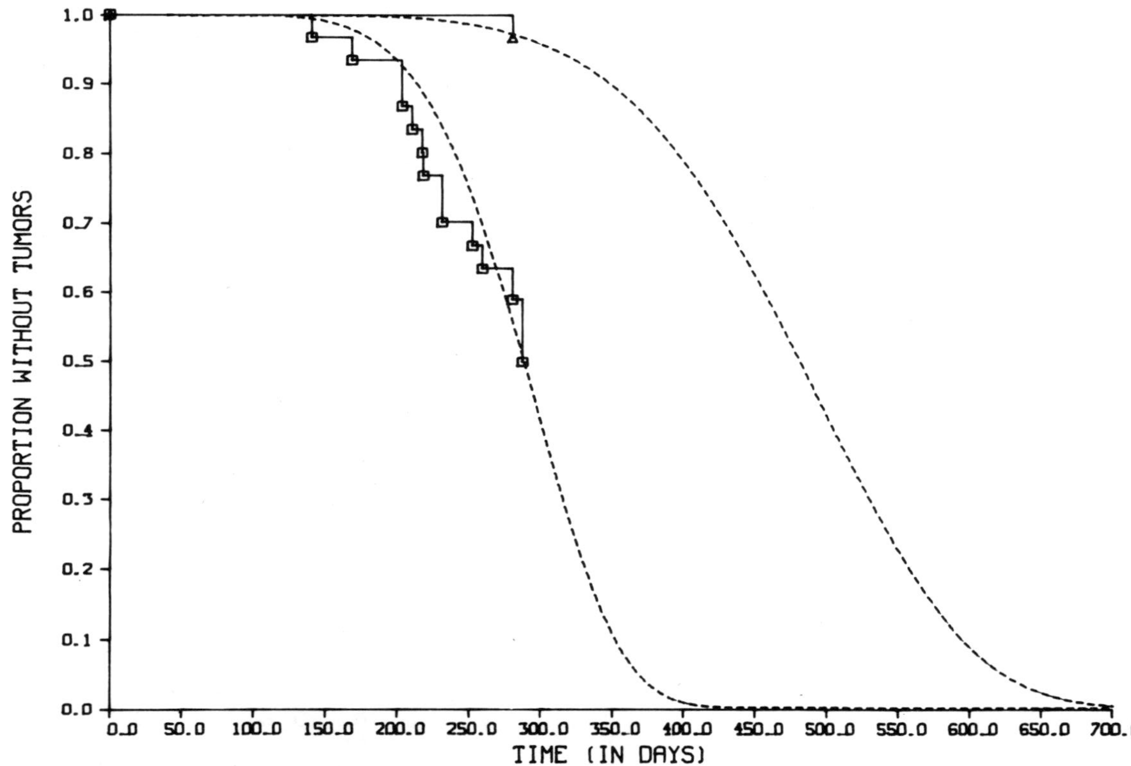

Figure 3. Proportion of mice surviving without skin tumor following application of the hydrotreated shale oil on a three times weekly schedule at 75 (□) mg per week and 50 mg (Δ) per week, twice weekly.

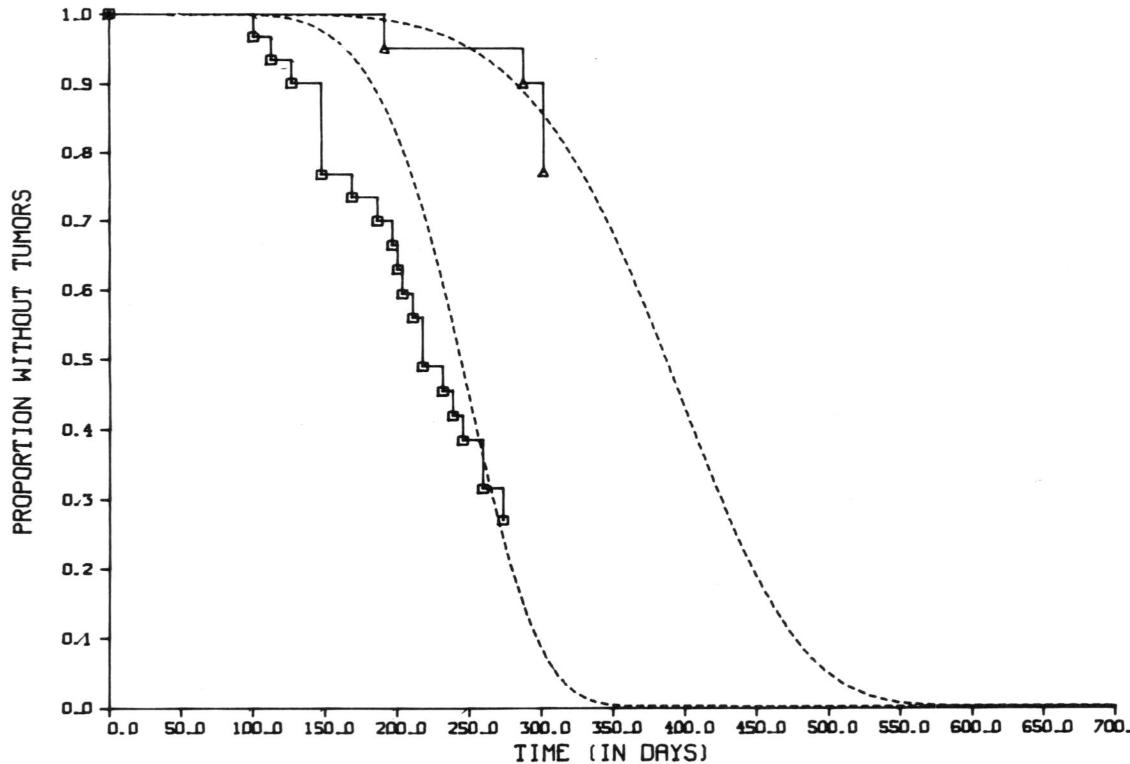

Figure 4. Proportions of mice surviving without skin tumor following application of the hydrotreated shale oil residue on a three times weekly schedule at 75 (□) mg per week and 50 mg (Δ) per week, twice weekly.

Table III. Relative Skin Carcinogenic Potencies of Shale Oil Crude, Hydrotreated Crude and Hydrotreated Residue on the Basis of 10 Months Exposure

Material	Dose Rate	Relative Potency	Upper 95% Confidence Limit
Shale oil crude	3/wk	1/998	1/425
Shale oil crude	2/wk	1/1698	1/713
HDT-Shale oil	3/wk	1/2780	1/1086
HDT-Shale oil	2/wk	1/8974	1/2851
HDT-Residue	3/wk	1/1622	1/664
HDT-Residue	2/wk	1/4634	1/1722

Non-Neoplastic Effects

To date, none of the distillates, whether derived from shale oil or natural petroleum, have induced skin neoplasms. However the materials have induced local inflammatory, degenerative and hyperplastic skin changes. The skin exposed to the test materials demonstrates loss of hair proportionate to the dosage level. At the highest exposure levels the surface hair loss is complete. The exposed skin is thickened, coarsely folded and deeply pigmented (Figure 5). Microscopically, the surface epidermis is both more cellular and covered by a thick layer of amorphous keratin (Figure 6). Small and large defects in the epidermis (ulcers) are present with attendant acute inflammation in the exposed dermis and the wound margins.

Hair follicles are always prominent features in chemically damaged mouse skin. In this particular instance hair follicles are exhibiting abortive hyperplasia with production of incompletely keratinized hair. Many branched and otherwise malformed follicles are observed deep within a generally disorganized and edematous dermis. The dermis is notable in that there are increased numbers of pigment cells clustered at the dermal-subcutaneous junction. These focal aggregates represent both phagocytized melanin arising from degenerating hair follicles as well as melanocyte proliferation. The upper portion of the follicle is filled with keratinaceous debris, hair fragments and occasional inflammatory cells.

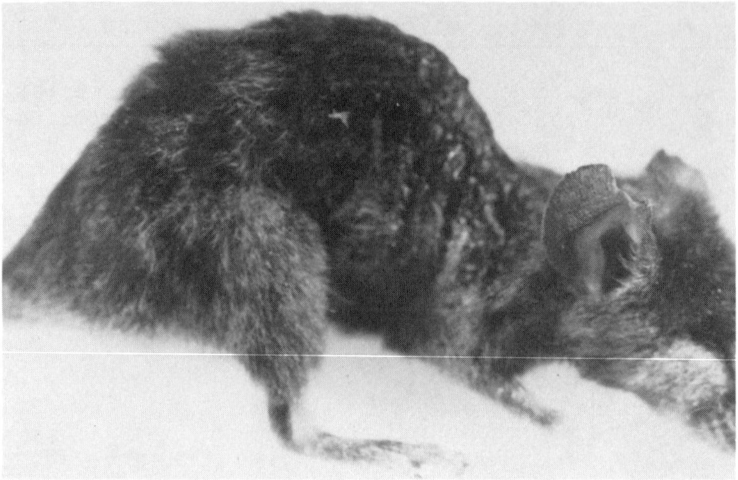

Figure 5. This mouse was one of two found dead in an upright and normal position in a group that had been exposed chronically to natural petroleum derived JP-5 (100% three times weekly). The exposed skin over the dorsal thorax is thickened, scaling and deeply pigmented.

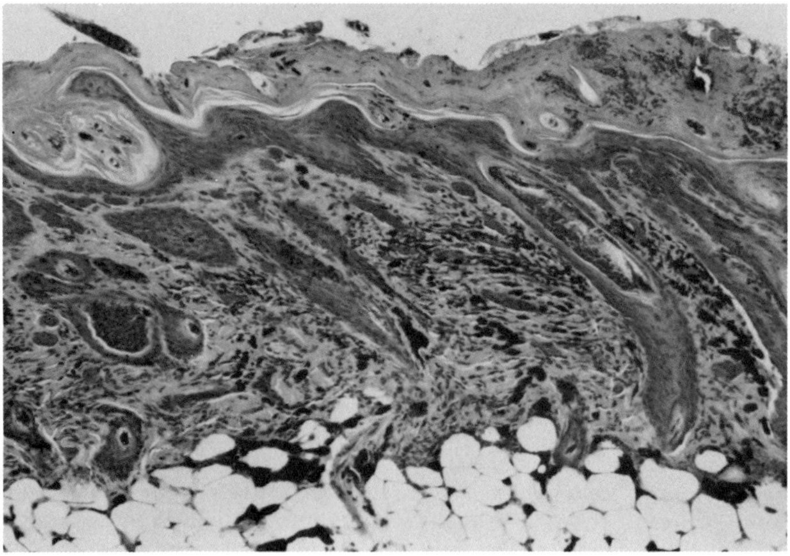

Figure 6. A photomicrograph of the skin of this mouse reveals an excess of amorphous keratin covering a hyperplastic epidermis. Hair follicles are also hyperplastic and sebaceous glands absent. The follicle lumina are filled with keratin, hair fragments and, occasionally, collections of neutrophil leukocytes. The dermis contains scattered aggregates of melanin which also is observed at the dermal-subcutaneous boundary.

The epithelium of the external root sheath has become continuous with the surface epidermis and in most areas the keratinization pattern noted on the surface is of the trichohyaline type. This indicates that epidermal keratinocytes have been overgrown by hyperstimulated follicular keratinocytes with attendant alteration in the type of keratin present on the surface.[5]

Marked ectasia of the superficial blood vessels is also noted in the dermis (Figure 7). In addition to the pigment aggregates mentioned previously, the dermis is infiltrated by variable numbers of mononuclear and neutrophil leukocytes. In summary, the skin of mice exposed to the undiluted distillates manifests a partially compensated chemical dermatitis with focal ulcerative and diffuse hyperplastic changes.

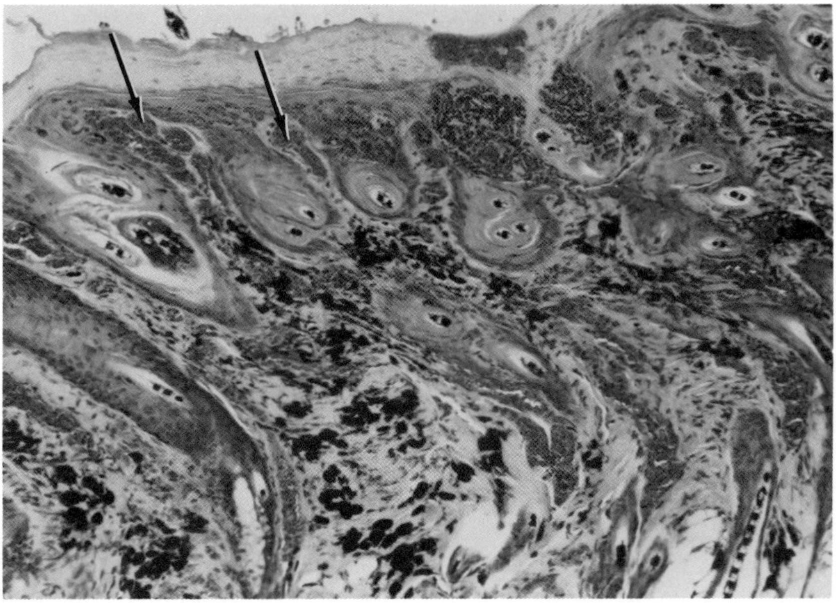

Figure 7. Dilated and congested capillaries are noted in the superficial dermis (arrows). Intraepithelial foci of inflammatory cells are also apparent.

In addition to local irritant effects the distillate materials have demonstrated systemic toxicity reflected both by a reduction in average body weight in groups exposed to the highest concentration of shale oil derived JP8 and DFM (Table IV), and an increase in average daily water intake observed in all groups exposed to the distillates but not the crudes (Table V).

The rather remarkable treatment and dose related increase in average daily water intake was noted commencing at about the eighth month of exposure and has continued through the present time without marked change. That the water consumption effect was treatment related was reinforced by the observation that when animals were not treated over the Memorial Day holiday water intake returned to control levels. When treatment was reinstituted after the holiday the water intake again increased to levels similar to those noted before the interruption in treatment schedule.

Table IV. Average Body Weight of Mice Exposed to Shale Oils and Distillates 3 Times Weekly for a Minimum of 30 Weeks

Material	Concentration (w/v %)	Average Body Wt. (gms. ± S.E.)
Shale oil crude	50	31.6 ± 0.9
HDT Shale oil	50	31.8 ± 0.9
HDT Residue	50	31.9 ± 1.3
JP5 Product	100	30.6 ± 0.5
JP8 Product	100	28.7 ± 1.0[a]
DFM Product	100	29.6 ± 0.9
Petroleum JP5	100	30.3 ± 0.6
Petroleum JP8	100	30.3 ± 0.7
Petroleum DFM	100	30.4 ± 0.7
Cyclohexane	100	31.5 ± 0.8

[a]Significant at the 0.05 level.

Increased water intake can occur as a consequence of pituitary/hypothalmic imbalances, kidney damage, changes in plasma osmolality such as accompany diabetes mellitus, and fluid losses that accompany severe burns, diarrhea, vomiting or external hemorrhage. Of these the most likely possibility was kidney damage arising from exposure to percutaneously absorbed materials or reactive metabolites generated by the

Table V. Average Daily Water Consumption in Mice Exposed to Shale Oils and Distillates for a Minimum of 30 Weeks

	3/week Exposures		2/week Exposures			
	Concentrations:		Concentrations:			
Material	100	50% (w/v)	100	50	25	12% (w/v)
Shale oil crude	nt[a]	nt	-	5.1(0.3)	4.9(0.1)	5.1(0.1)
HDT Shale oil	nt	nt	-	5.0(0.1)	4.7(0.2)	4.7(0.2)
HDT Residue	nt	nt	-	4.8(0.3)	4.0(0.1)	4.8(0.3)
JP5 Product	9.9(0.3)[b,c]	7.3(0.3)	8.3(0.3)	6.6(0.1)	5.7(0.3)	4.7(0.1)
JP8 Product	9.4(0.4)[c]	7.3(0.3)	8.2(0.2)	6.7(0.3)	5.7(0.3)	5.1(0.2)
DFM Product	10.9(0.3)[c]	7.8(0.2)	9.2(0.4)	6.9(0.2)	6.1(0.3)	5.0(0.1)
Petroleum JP5	10.4(0.4)[c]	7.3(0.2)	8.1(0.1)	6.7(0.3)	5.1(0.1)	4.4(0.1)
Petroleum JP8	10.2(0.3)[c]	7.5(0.1)	8.2(0.1)	6.8(0.1)	5.4(0.3)	4.9(0.2)
Petroleum DFM	10.2(0.3)[c]	7.7(0.2)	9.3(0.2)	6.8(0.2)	5.4(0.1)	5.3
Cyclohexane	5.1(0.1)	-	-	-	-	-

[a]Not tested. [b]Average daily water intake in ml/mouse/day ($\pm$ S.E.).
[c]Significant at the 0.001 level.

hyperplastic skin. To evaluate this possibility individual mice were placed in a metabolism cage in which water intake, urine production and feces production can be accurately monitored. The availability of only one cage has limited the amount of data we were able to collect for this presentation but the limited data we have obtained (Table VI) reveal an interesting phenomenon. We have three independent measurements of mice treated with cyclohexane (solvent) three measurements of mice treated with shale oil derived DFM (DFMS), and two measurements in mice treated with natural petroleum derived DFM. These data confirmed the higher average daily intake in mice exposed to distillates relative to vehicle controls but this greater water intake cannot be explained as a consequence of loss of renal concentrating ability because urine production rates were only slightly above control. The only other possible sources of water loss was through feces, percutaneous diffusion and the respiratory tract. Fecal weights did not differ between treated and control. By a process of elimination this leaves the respiratory tract and precutaneous diffusion as the remaining alternatives. Of these, water loss through damaged skin is the more likely possibility. This hypothesis is ammenable to experimental test by treating the damaged skin in a way to reduce its permeability to water molecules, while continuously monitoring the water intake and output profiles.

The possibility that a fraction of the increased water requirement was a consequence of kidney damage is suggested by histologic finding in two animals dying acutely in the group exposed to 100% natural petroleum derived JP5. The tissues from these animals were subjected to histologic evaluation. The changes in the skin were described previously. The liver exhibited lobular atrophy and focal coagulation necrosis. There was arrest of follicular maturation in the ovary with persistence of corpora lutea and concomitant cystic hyperplasia of the uterus and mammary tissue; the mice appeared pseudopregnant. The adrenals exhibited mild cortical atrophy and the thyroid and pituitary were unremarkable.

The kidneys were grossly normal. Microscopic lesions were observed at the corticomedulary junction. The lesions were restricted to the terminal straight segment of the proximal tubule and the thick ascending limb and distal tubule. The glomerulus, convoluted proximal tubule and collecting ducts were not involved. The lesions consisted of patchy necrosis of tubular epithelial cells with segmental loss of epithelium (Figure 8). Affected tubules were filled with pink amorphous material probably representing cytoplasmic debris from the disintegrating epithelium. Pycnotic nuclei were common in the affected tubules. The lesions,

Table VI. Water Balance in Mice Exposed Chronically to Petroleum and Shale Oil Derived Middle Distillates

Treatment	Determinations	Average Water Intake (ml)	Average Urine Output (ml)	Average Fecal Weight (gm)
Cyclohexane	3	3.7	1.3	1.1
DFM Product	3	7.7	1.6	1.3
Petroleum DFM	2	9.0	2.4	1.5

while easily recognized were not considered to be either sufficiently extensive or long standing to explain increased water consumption. Both mice died per-acutely, literally "in their tracks," (see Fig. 5). This fact, coupled with other anatomic indications of a shock syndrome (contracted spleen, acute ischemic myocardial necrosis) make it likely that the mice died as a consequence of water deprivation. Dehydration can produce focal tubular necrosis similar in character, degree and distribution to the lesions described. Sequential sacrifice and perfusion fixation of mice from the twice weekly exposure group will clarify the frequency and extent of this lesion.

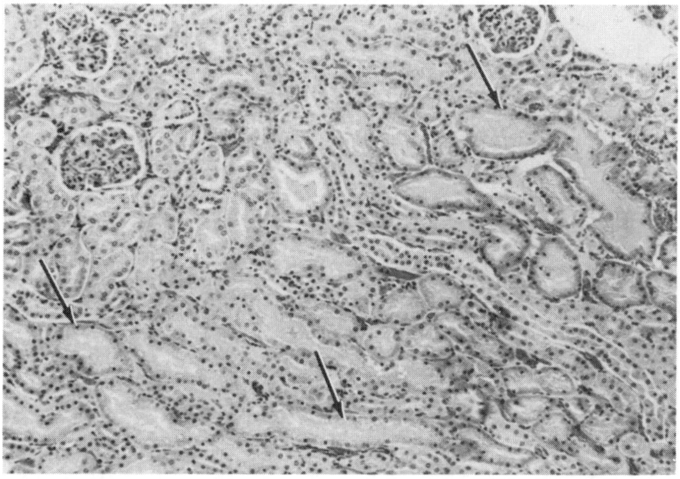

Figure 8. Renal corticomedullary junction. Tubular epithelium was undergoing necrosis as reflected by an increased staining intensity and disintegration. Nuclei in affected tubule segments were infrequent and often pyknotic. Degenerating tubules were also dilated and occluded by cellular debris (arrows).

In spite of the possible contribution of skin damage to water loss and the potential role of dehydration in the pathogenesis of renal lesions noted in this study, similar renal lesions were noted by Carpenter et al.[6] in rats exposed by inhalation to certain petroleum solvents. Human contact exposures to diesel fuel have also been associated with acute renal failure on at least two occasions.[7,8] These observations, combined with the present experience, increase the likelihood that "middle distillates" are, at least potentially, nephrotoxic. The basis for nephrotoxicity may vary depending upon the route of exposure, dose rate and chronicity. If exposures are massive, direct effects are more likely. Alternatively, chronic dermal exposures may sufficiently alter skin permeability both to body water as well as toxic materials. In an animal as small as the mouse, a small percentage change in skin permeability could have a large effect upon water requirements given the proportionately high surface to volume ratios involved.

Several reports have appeared that describe the acute toxicity of petroleum and shale oil derived military fuels. Smith et al. (this volume) report a low level of systemic toxicity. Cowan et al.[9] reported LD_{50}'s for petroleum and shale oil derived JP5 and DFM. These authors reported that shale oil products were slightly more toxic than petroleum derived materials. JP5 inhalation for 120 days caused no unusual hematologic or histopathologic changes in rats and guinea pigs. Neurobehavioral effects have also been identified in a case-control study of workers exposed to petroleum base jet fuel.[10]

In a study of the chronic toxicity of JP4 jet fuel, Kinkead et al.[11] reported no marked toxic effect in mice, rats, dogs and monkeys exposed repeatedly to 2.5 or 5.0 mg JP-4 vapor/L for 6 hours/day, 5 days/week. However, a slight increase in tumor incidence in mice, activity depression in dogs and monkeys, enhanced erythrocyte fragility in dogs and increased organ weights in the rat led the authors to recommend an occupational exposure limit not to exceed 2.5 mg vapor/L.

In summary, it would appear that the principal toxiologic concern associated with chronic dermal exposure to Paraho Shale Oil Crude, the Hydrotreated Crude and the Hydrotreated Residue is skin tumor induction. For distillates in the boiling range of those in the current study the probability of skin tumor induction is very low; however, the potential for local irritation and systemic toxicity is greater than with the crude materials.

REFERENCES

1. M. A. Pike, "A Suggested Method of Analysis of a Certain Class of Experiments in Carcinogenesis." Biometrics 22, 142-161 (1966).

2. R. Peto and P. N. Lee, "Weibull Distributions for Continuous-Carcinogenesis Experiment." Biometrics 29, 457-470 (1973).

3. R. Peto, P. N. Lee, and W. S. Paige, "Statistical Analysis of the Bioassay of Continuous Carcinogens," Br. J. Cancer 26, 258-261 (1972).

4. J. M. Holland, D. G. Gosslee and N. J. Williams, "Epidermal Carcinogenicity of Bis(2,3-epoxycyclopentyl)-ether, 2,2-Bis(p-glycidyloxyphenyl)propane, and m-Phenylenediamine in Male and Female C3H and C57BL/6 Mice," Cancer Res. 39, 1718-1725 (1978).

5. H. Pinkus, "Keratosis Senilis: A Biologic Concept of Its Pathogenesis and Diagnosis Based on the Study of Normal Epidermis and 1730 Seborrheic and Senile Keratoses," Am. J. Clin. Path. 29, 193-207 (1958).

6. C. P. Carpenter, E. R. Kinkead, D. L. Geary, L. J. Sullivan and J. M. King, "Petroleum Hydrocarbon Toxicity Studies. VI. Animal and Human Responses to Vapors of '60 Solvent'," Toxicol. and Appl. Pharmacol. 34, 374-394 (1975).

7. A. M. T. Barrientos. T. Ortuno, J. M. Morales, F. Martinez-Tello and J. L. Rodicio, "Acute Renal Failure After Use of Diesel Fuel as Shampoo," Arch. Intern. Med. 137, 1217 (1977).

8. A. J. Crisp, A. K. Bhalla and B. I. Hoffbrand, "Acute Tubular Necrosis After Exposure to Diesel Oil," British Med. J. 2, 177 (1979).

9. M. J. Cowan, L. J. Jenkins, Jr., L. E. Doptis and R. W. Young, "Evaluation of Comparative Toxicity and Shipboard Hazards of Selected Petroleum and Oil-Shale-Derived Fuels," Oil Shale Symposium Proceedings Vol. II, 1978, pp. 21-31.

10. B. Knave, B. A. Olson, S. Elofsson, F. Gamberale, A. Isaksson, P. Mindus, H. E. Persson, G. Struwe, A. Weinberg and P. Westerholm, "Long Term Exposure to Jet Fuel. II. A Cross-Sectional Epidiologic Investigation on Occupationally Exposed Industrial Workers with Special Reference to the Nervous System," Scand. J. Work Environ. Hlth. 4, 19-45 (1978).

11. E. R. Kinkead, L. C. DiPasquale, E. H. Vernot and J. D. MacEwen, "Chronic Toxicity of JP-4 Jet Fuel," Proceedings of 5th Annual Conf. Environ. Toxicol., ISS AMRL-TR-74-125, 1974, pp. 145-154.

CHAPTER 8

COMPARATIVE DERMOTOXICITY OF SHALE OILS

L. M. Holland, J. S. Wilson, and M. E. Foreman, Toxicology Group, Los Alamos National Scientific Laboratory, Los Alamos, New Mexico 87544

INTRODUCTION

Many liquid fossil fuels have been shown to be carcinogenic after repeated application to the skin. Historically the shale oils have been particularly associated with occupational diseases of the skin. While most attention has been given to neoplastic phenomena, a considerable array of acute and chronic dermatoses have been reported.[1] A sex-related difference in either tumor occurrence or inflamatory response has not been frequently reported in the experimental literature, however an increase in skin cancer among female Estonian oil shale workers has been observed.[2] In our experience when shale oils are applied at higher dose levels the standard observation of tumor production and latency are often obscured by a severe inflammatory response leading to epidermal degeneration. The two experiments reported here are still in progress, however the interim results are useful in assessing both the phlogistic and tumorigenic properties of three shale oils.

EXPERIMENTAL APPROACH

Three shale oils were tested in these experiments. The first crude oil (OCSO #6) was produced in a modified *in situ* retort at Occidental Oil Company's Logan Wash site near Debeque, Colorado. The second crude oil (PCSO II, Repository No. 4601) was produced in the above ground Paraho vertical-kiln retort located at Anvil Points near Rifle, Colorado and the third oil was the hydrotreated daughter product of the Paraho crude (PCSO-UP, Repository No. 4602). Both Paraho oils were obtained from the EPA/DOE Repository at Oak Ridge National Laboratory. The Occidental oil was obtained from the DOE Oil Shale Task Force sample-suite associatd with Logan Wash retort #6. Each oil was diluted using a mix of cyclohexane and acetone (30% cyclohexane/70% acetone) as the vehicle. A total volume of 50 μl (oil plus vehicle) was used for each dose in all experiments.

In both experiments equal numbers of male and female C3Hf/He mice obtained from the animal colonies at Oak Ridge National Laboratory were used at each dose level. All mice were started in the experiments at 10-12 weeks of age. The hair in the intrascapular area of each mouse was clipped prior to the first treatment and as needed thereafter.

Experiment I was designed to determine the highest dose level at which tumor latency could be measured without interference from epidermal degeneration (Table I). The 10 mg dose group was started later after experience with higher doses identified severe cytotoxicity. A 3 times weekly dose frequency was used at all levels.

Experiment II was designed to determine the effect of application frequency on both tumor response and inflammatory phenomena (Table II). Complete epidermal degeneration was used as the only measure of severe inflammation. Relative tumorigenicity was based on the number of tumor bearing mice without regard to multiple tumors on individual animals. In both experiments, tumor occurrence was confirmed one week after initial appearance.

One group of animals, treated 3 times/week with the solvent vehicle, was used as a control for both experiments and since no overt inflammatory responses had been observed in animals treated with natural petroleum during previous experiments none were tested.

DISCUSSION

The results to date of the first experiment are included in Table I. The effect of cytotoxicity is apparent at both the 20 mg and 40 mg dose levels. At the 20 mg level the PCSO II crude oil was less destructive of tissue but more carcinogenic. The modified <u>in situ</u> oil, OCSO #6, caused fewer tumors but because of cytotoxicity also reduced the number of cells at risk. At the highest dose level both crude oils produced the same number of tumor bearing mice after 33 weeks of exposure but the high levels of both cytotoxicity and animal lethality obscure any analysis of tumor occurrence. The data show a pronounced difference in response between the two crude oils and the hydrotreated Paraho oil. While the upgraded oil is clearly dermotoxic at high doses, the observed tumor incidence and animal lethality is lower than either of the crude oils. This diminished tumorigenicity may be associated with a reduction in total nitrogen content.[3]

Table I

Oil	(Repository No.)	Dose[a]	No. with Tumors	No. with ED[b]	No. Dead	No. Days
PCSO II	(4601)	5 mg	10	0	0	238
PCSO-UP	(4602)	5 mg	1	0	0	238
OCSO #6	--	5 mg	7	2	0	231
PCSO II	(4601)	10 mg	2	0	0	85
PCSO-UP	(4602)	10 mg	0	0	0	85
OCSO #6	--	10 mg	1	1	0	85
PCSO II	(4601)	20 mg	14	6	4	238
PCSO-UP	(4602)	20 mg	0	19	3	238
OCSO #6	--	20 mg	6	30	9	231
PCSO II	(4601)	40 mg	12	12	14	238
PCSO-UP	(4602)	40 mg	2	8	2	238
OCSO #6		40 mg	12	24	24	231

[a]40 mice (20 male/20 female) per dose level
[b]Complete epidermal degneration

Table II

Oil	(Repository No.)	Dose/Frequency[a]	ED		Tumors[b]	
			Males	Females	Males	Females
OSCO #6	--	10 mg x 4 (40 mg)	9	8	1	0
		20 mg x 2 (40 mg)	0	0	1	0
		40 mg x 1 (40 mg)	0	0	0	0
		20 mg x 1 (20 mg)	0	0	0	0
		20 mg x 2 (40 mg)	0	0	1	0
		20 mg x 3 (60 mg)	10	5	2	0
PCSO II	(4601)	10 mg x 4 (40 mg)	7	0	1	0
		20 mg x 2 (40 mg)	1	0	1	0
		40 mg x 1 (40 mg)	0	0	0	0
		20 mg x 1 (20 mg)	0	0	0	0
		20 mg x 2 (40 mg)	1	0	1	0
		20 mg x 3 (60 mg)	8	2	2	0

[a]Frequency/week - 20 mice (10 male/10 female) per dose
[b]After 18 weeks of treatment

The Paraho crude oil, PCSO II, contains 2.21% nitrogen and the hydrotreated daughter oil, PCSO-UP contains 0.55% nitrogen. The other crude oil, OCSO #6, has a nitrogen content (1.49%) which falls about midway between that of the two Paraho oils, but is closer to PCSO II in potency. Nitrogen content, while it may act as an indicator of carcinogenic potency, is a crude measure at best and it is probable that a whole array of compounds, including some nitrogen containing ones, play a role.

A general inflammatory response by the epidermis is an important factor in experiments designed to study carcinogenesis. Both total tumor production and latency are affected by cell killing which reduces the effective cells at risk. Without very large dose groups, small differences in dose response may be difficult to detect in the presence of significant inflammation. The observed animal lethality associated with the more inflammatory high doses has not been defined. Clinically, there appeared to be an increase in both water intake and urination by affected animals. Whether lethality is related to significant loss of epidermis and dehydration or to a direct renal toxicity has not been determined.

The second experiment, dealing with application frequency, indicates a strong influence related to fractionation of a given total dose within a fixed time period (Table II). These data indicate that the number of times an oil is applied has profound effect on its phlogistic potential. With both oils, a single 40 mg dose applied once a week was not overtly cytotoxic while the same total dose divided into four equal applications had a severe cell killing effect. Repair phenomena, particularly the temporal aspects of cell replacement, are undoubtedly the main mechanisms involved. The experiment has not yet defined a relationship between dose frequency and tumor latency. The sex-related difference in inflammatory response is striking and certainly has significance for experimental design. An increased phlogistic sensitivity expressed in male mice could affect the meaning of an experiment where only one sex was used. It is not clear at this time if this sex-related difference extends to total tumor occurrence and latency.

The tumorigenic potential of shale oil has been fairly well established and at least one effective mitigating procedure has been identified. However some important facets of the problem remain unanswered. Benign dermatoses can be significant occupational health problems and more data dealing with the phlogistic properties of the shale oils are needed if we are to provide meaningful information for worker protection guidelines. We must also determine if there are process-specific differences between oils and attempt to define the factors involved. This potential process-specificity may be important to both the tumorigenic and the phlogistic aspects of occupational skin disease.

REFERENCES

1. A. Scott, "On the Occupation Cancer of the Paraffin and Oil Workers of the Scottish Oil Shale Industry," Brit. Med. J., December 1922.

2. M. Purde and M. Rahu, "Cancer Patterns in the Oil Shale Area of the Estonian S.S.R.," Environ. Health Pers. 30 (1979).

3. R. M. Coomes, "Carcinogenic Aspects of Oil Shale," in Proceedings of Environmental Aspects of Non-Conventional Energy Resources, Denver, Colorado, 1978 (American Nuclear Society Topical Meeting).

CHAPTER 9

CARCINOGENIC BIOASSAY OF SHALE OIL REFINERY STREAMS AND DOWNSTREAM PRODUCTS

S. C. Lewis, Exxon Corporation, East Millstone, NJ 08873

INTRODUCTION

Crude and upgraded shale oils have been shown to be carcinogenic in laboratory animals.[1,2,3] Massive exposure under unhygienic conditions has been associated with increased incidence of skin cancer in man.[4] The carcinogenic potencies of shale oils have been reported to vary widely.[5,6,7] These potencies vary as a function both of the temperature at which the raw shale oil is retorted and of the severity of the refining/upgrading of the retorted oil. The influence of shale source and other potential contributions to variable biological activity have not been rigorously studied.

MATERIALS AND METHODS

In the present studies sponsored by the American Petroleum Institute (API), classical "lifetime mouse dermal carcinogenesis bioassay experiments" were carried out on shale-derived materials. The exact list of samples tested, with various identifiers, is shown in Table I; more detailed information on the specifications of these samples is available elsewhere (See Chapter 4).[8] Samples AP-79-3 through -7 were made available to API through the cooperation of the United States Navy and the Standard Oil Company of Ohio (SOHIO). The crude shale oil (AP-79-3) was prepared from raw northwest Colorado shale by the PARAHO process, and samples AP-79-4, -5, -6, and -7 were, in turn, prepared from that crude supply. The insitu process shale oil was manufactured in a true underground retort; details of the sample identity and process history are available from API*. Also shown in Table I are details regarding the negative control, solvent vehicle control and positive control bioassays which were conducted in concert with the shale oil experiments.

---

*Contact D. B. Disbennett, American Petroleum Institute, 2101 L Street, N.W., Washington, DC 20037, (202) 457-7192.

Table I. Test Sample Identification

Test Abbreviation Designation	Description	Repository No.	Code
AP-79-3	Retort Oil (Crude Shale Oil)	4601	RO
AP-79-4	Hydrotreated Shale Oil	4602	HSO
AP-79-5	Diesel Fuel Marine, Precursor	4606	DFM/pre
AP-79-7	Diesel Fuel Marine, Product	4610	DFM/pro
AP-79-6	Hydrotreated Residuum	4607	HR
AP-79-8	In-Situ Shale Oil	--	ISO
AP-79-1	No Chemical Treatment	--	NT
AP-79-2	Toluene (Solvent/Vehicle)	--	SV
AP-79-9	0.05% BaP (in Toluene)	--	PC-LD
AP-79-10	0.15% BaP (in Toluene)	--	PC-HD

A comprehensive matrix of experiments was conducted on each of the shale oil samples. In addition to the bioassay for carcinogenic activity, each sample was submitted for extensive physical/chemical characterization (unpublished data available on request from API). Finally, state-of-the-art chemical analyses of all shale oil samples for aggregate and individual polynuclear aromatic hydrocarbons were performed (See Chapter 5).[9] Note: Due to the very preliminary status of these cancer bioassays, no attempt will be made here to correlate the results of the chemical/analytical and biological experiments.

Figure 1 shows the salient elements of the animal husbandry for these studies; Figure 2 shows the equivalent detail for treatment and observation of animals under the contractor's standard protocol (available from Kettering Institute, University of Cincinnati upon request**).

RESULTS

Table II shows the interim results reported as of the end of the 30th week of continuing studies. At this point in the study, only the diesel-fuel-marine-precursor and product have failed to produce tumors, excepting the negative control. Both negative and positive control experiments are

**Contact W. B. Barkley, Department of Environmental Health, College of Medicine, University of Cincinnati, Cincinnati, OH.

progressing as have similar studies in this and other laboratories.[10,11] Tumor incidence data are shown in the third column of Table II. Tumors are classified soley on the basis of their gross characteristics; neither "benign" nor "advanced" tumors (as reported in column 4) have been examined microscopically. Therefore, these data must be considered as <u>tentative</u> tumor classifications. Specific histopathologic examinations will be available only after completion of the experiment.

- Weanling C3H/HeJ male mice
- Two week quarantine (with rejection of morbid animals)
- Unique identification by toe-clip
- Gang Housing in groups of 5
- Standard chow and water <u>ad lib</u>
- 50 animals/group -- all groups (including controls)
- Interscapular area clipped free of hair, p.r.n.
- 50 mg/application -- calibrated pipette
- 100% concentration (unless specified otherwise)
- 2 applications/week
- 18 months (or) appearance of carcinoma (gross clinical diagnosis)
- Twice-monthly body weight determination
- Complete gross necropsy (per NCI suggested procedures)
- Histopathologic examination (formalin fixation)
    - Skin from treatment site (<u>all</u> animals)
    - Grossly observed tumors (regardless of site)
    - Any suspected lesion
- Monthly reports of tumor status/progress

Figure 1. Protocol.

Table III shows the average tumor latencies for animals from groups treated with the retort oil, hydrotreated oil, hydrotreated residue, in-situ oil, and benzo(a)pyrene. The average tumor latency is computed by summation of the individual elapsed times (in weeks) for the appearance of the <u>first</u> tumor in each tumor-bearing animal divided by the total number of tumor-bearing mice (TBM). The total numbers of TBM's upon which the mean latencies are calculated are shown in the last column of Table III. A relatively shorter apparent latencies of the tumors associated with the hydrotreated shale oil (HSO) and low dose positive control may be statistical artifacts of the low total numbers of tumorbearing mice on which these means were calculated.

Table II. Results[a] (As of 30th Week of Study)

Sample	Repository No.	Survivors Without Tumor	Total Number With Benign Tumor(s)	Total Number With Advanced Tumor(s)
RO	4601	2	22	20
HSO	4602	39	4	2
DFM/Pre	4606	49	0	0
HR	4607	30	13	5
ISO	--	0	12	37
NT	--	50	0	0
SV	--	47	0	0
PC-LD	--	45	3	2
PC-HD	--	9	20	21

[a]Expressed as tumor bearing animals; original "n" = 50

Table III. Results (As of 30th Week of Study)

Sample	Repository No.	Average Tumor Latency[a]	"n"[b]
RO	4601	20.0	42
HSO	4602	23.2	6
HR	4607	28.4	18
ISO	--	14.4	49
PC-LD	--	13.4	5
PC-HD	--	25.8	41

[a]Expressed in weeks
[b]"n" = Sum of all animals bearing tumors, benign or advanced; divisor for computing average tumor latency

CONCLUSIONS AND DISCUSSION

These are interim data from studies in progress at the time of publication and as such, conclusions, in the true sense of the word, are inappropriate. However, some preliminary observations can be offered. The total incidence of tumors among mice treated with the raw shale oil (42 of 50 mice bearing tumors) and the in-situ shale oil (49 of 50 mice bearing tumors) express very high tumorigenic potencies, comparable to the high dose positive control. The somewhat lower tumor incidences, 12 and 36 percent among mice treated with the hydrotreated shale oil and hydrotreated residue, respectively, may be assumed to reflect more moderate biological activities. Since the positive and negative control studies appear to be following courses essentially as have similar experiments by other investigators, one may assume the validity of the animal model has been reconfirmed. On the basis of these observations, one may tentatively classify a gradient of tumor activities as follows:

ISO>RO>HR>HSO>DFM's (currently inactive).

REFERENCES

1. J. M. Holland, D. A. Wolf, and B. R. Clark, "Relative Potency Estimation for Synthetic Petroleum Skin Carcinogens," Environmental Health Perspectives, in press.

2. J. M. Holland, "Experimental Evaluation of Skin Carcinogenicity Associated with Chronic Exposure to Synthetic Petroleum," Proceedings of Toxicology Forum., July 27-August 1, 1980, Aspen, Colorado.

3. C. C. Twort and J. M. Twort, "The Carcinogenic Potency of Mineral Oil," J. Indust. Hyg., 13, 204-226 (1931).

4. A. Scott, "The Occupational Dermatoses of the Paraffin Workers of the Scottish Shale Oil Industry," Eighth Scientific Report of the Imperial Cancer Fund, Taylor & Frances, London, 1923, pp. 85-142.

5. W. E. Smith, R. A. Sunderland, and K. Sugiura, "Experimental Analysis of the Carcinogenic Activity of Certain Petroleum Products," Arch. Indus. Hyg., 4, 299-314 (1951).

6. J. M. Holland, L. C. Gipson, M. J. Whitaker, T. J. Stephens, G. M. Clemmer, and R. A. Wolf, "Chronic Dermal Toxicity of PARAHO Shale Oil and Distillates," unpublished.

7. J. M. Holland, "Dermal Effects of PARAHO Shale Oil and Derived Distillates," Proceedings of the "Symposium on Health Effects Investigation of Shale Oil Development," Gatlinburg, TN, June 23-24, 1980.

8. W. H. Griest, M. R. Guerin, L. B. Yeatts, and B. R. Clark, "Sample Management and Chemical Characterization of the PARAHO/SOHIO/U.S. Navy Crude and Refined Shale Oil Suite," Proceedings of the "Symposium on Health Effects Investigation of Shale Oil Development," Gatlinburg, TN, June 23-24, 1980.

9. W. K. Robbins, and S. C. Blum, "Analysis of Shale Oil Liquids for Polynuclear Aromatic Hydrocarbons and Their N-Heterocyclic Analogs," Proceedings of the "Symposium on Health Effects Investigation of Shale Oil Development," Gatlinburg, TN, June 23-24, 1980.

10. J. M. Holland, B. M. Eisenhower, L. C. Gipson, L. H. Smith, T. J. Stephens, and M. S. Whitaker, "Assessment of the Carcinogenic Risk from Energy Related Organics," presented at Fourth National Conference, Interagency Energy/Environment R&D Program, Washington, D.C., June 7-8, 1980.

11. C. S. Weil, R. M. Coomes, Barkley, W., Holland, J. M., and S. C. Lewis, personal communications of unpublished data.

CHAPTER 10

NAVY TOXICITY STUDY OF SHALE AND PETROLEUM JP-5 AVIATION FUEL AND DIESEL FUEL MARINE*

LCDR Morris J. Cowan, Jr., MSC, USN, CDR Lawrence J. Jenkins, Jr., MSC, USN, Naval Medical Research Institute Toxicology Detachment (NMRI/TD), Wright-Patterson AFB, Ohio 45433

INTRODUCTION

The Navy's principle interest in nonpetroleum fossil fuel sources, or synthetic fuels, has been to determine their compatibility with naval equipment and engines, and their suitability as naval fuels. Suitability includes engineering and health effects. Current interest in synthetic fuels remains focused on crude shale oil from surface retorting, and the Navy has participated in an interagency effort to produce and refine a large quantity of crude shale oil into Military Specification fuels for subsequent test and evaluation.[1] Recently these efforts resulted in the refinement by Sohio of 80,000 bbl of Paraho shale crude. In December of 1978 shale derived JP-5 grade aviation turbine fuel and diesel fuel marine, or DFM, were made available to the Naval Medical Research Institute/Toxicology Detachment for toxicology testing.

In order to assess the suitability of shale JP-5 and DFM, it was important that the toxicity of the fuels be

---

*Naval Medical Research and Development Command, Research Task No. MF65572001.4009. The opinions and assertions contained herein are the private ones of the writer and are not to be construed as official or reflecting the views of the Navy Department or the Naval Service at large. The experiments conducted herein were conducted according to the principles set forth in the current edition of the "Guide for the Care and Use of Laboratory Animals," Institute of Laboratory Animal Resources, National Research Council.

characterized keeping in mind the potential personnel exposure hazards ashore and aboard ship. Comparative data on the toxicity of the synthetic and petroleum fuels were necessary to assess engineering controls and work procedures. Unfortunately, little data was available on the dermal, occular or inhalation toxicity of the distillate fuels used by the Navy; thus, precluding a comparative evaluation of the presently used petroleum fuels versus shale fuel. The lack of such data prompted the series of investigations whose status I will report on today.

MATERIALS

First I would like to spend a few moments discussing the fuels used in these investigations. Table I shows the four fuels studied. They were shale and petroleum JP-5 aviation turbine fuel and shale and petroleum DFM. The military specifications allow for quite a bit of variation from one batch of fuel to another as is evident in Table II.

JP-5

JP-5 is a high flash point aviation turbine fuel with a specified distillation temperature of 205°C for the 10% recovery point to 290°C for the end point.[2] Specifications permit up to 25% aromatic hydrocarbons and 5% olefins. Substituted phenols or amines are added to inhibit oxidation.

DFM

Diesel fuel marine is also a distillate fuel, with a boiling range whose only limitation is that it be lower than 385°C.[3] The content of aromatics and olefins is not specified, but is limited only by the requirement that the fuel be clear and bright in appearance. As Table II indicates the fuel is even less well defined chemically than JP-5.

All four fuels met military specifications with minor exceptions. The shale JP-5 had 0.21 volume percent of fuel system icing inhibitor (ethylene glycol monomethyl ether) when the military specification allows only 0.15 volume percent, and it failed the copper strip corrosion test, a subjective test for "troublesome sulfur compounds", to quote the engineers in the fuels lab.

METHODS

I'll begin the discussion of toxicity studies performed with these four fuels by describing the efforts to identify ocular, dermal and sensitization potentials of the fuels. Comparative primary dermal irritation tests were performed

Table I. Fuels Comparative Toxicity Study

Fuel	Source
Petroleum JP-5	Air Force Aero Propulsion Laboratory, Wright-Patterson Air Force Base, Ohio
Shale JP-5	100,000 bbl Paraho Crude, Sohio Refinement (1978)
Petroleum DFM	Sohio
Shale DFM	100,000 bbl Paraho Crude, Sohio Refinement (1978)

Table II. Some Selected Military Specifications for JP-5 and DFM[a]

	JP-5	DFM
Distillation temp °C		
Initial B.P.	---	---
10% Recovery	205	---
End point, max temp	290	385
Aromatics, vol %, max	25	---
Olefins, vol %, max	5	---
Sulfur, total wt %, max	0.4	1.0
Sulfur, mercaptan, wt %	.001	---
Hydrogen content, wt%	13.5	---
Freezing point °C, max	- 46	---
Density, g/ml at 15°C, min	0.788	---
Density, g/ml at 15°C, max	0.845	---
Flashpoint, °C, min	60	60

[a]Source: MIL-T-5624K,Turbine Fuel, Aviation Grades JP-4 and JP-5, 1 April 1976.
MIL-T16884G, Fuel Oil, Diesel, Marine, 7 March 1973.

on all four fuels. The test was a patch test method conducted to measure the degree of primary dermal irritation of intact and abraded skin. Six rabbits were used for each fuel sample. The samples were in place for 24 hours. After removal of the sample patches, the test areas were scored according to the standard Draize technique.[4]

Primary eye irritation tests were performed using nine rabbits for each fuel. A 0.1 milliliter sample of fuel was applied to one eye of each rabbit. The opposite eye served as a control. Three of the treated eyes were flushed with water approximately 30 seconds after treatment with the fuel in order to assess the effectiveness of emergency treatment techniques in preventing eye injury. Irritative effects were scored at 24, 48 and 72 hours using the Draize technique. Dermal sensitization tests conducted on the four fuels were a modification of the Landsteiner technique[5] in that the size of the irritated area was evaluated as well as the intensity of the irritation. Twenty-four guinea pigs per fuel were given a total of seven sensitizing doses each of a 0.1 percent dilution of fuel in peanut oil over a two week period. The guinea pigs were untreated for three weeks and then given a challenge dose of fuel in peanut oil. Simultaneous injections of peanut oil alone were given as a control to each animal. Reaction to the challenge dose taking into consideration the peanut oil control reaction was recorded as a product of the length and width of the wheal multiplied by a numerical reaction score which described the reaction intensity in increasing severity.

The four major inhalation toxicity studies conducted are outlined in Table III. The studies are being conducted at the Naval Medical Research Institute, Toxicology Detachment at Wright-Patterson Air Force Base, Ohio, utilizing Thomas Dome inhalation chambers at the Air Force Aerospace Medical Research Laboratory's Toxic Hazards Research Unit. All four fuels were studied under identical protocols with the exception of different chamber vapor concentrations of the JP-5 fuels and the DFM fuels. Exposures were subchronic continuous 90-day fuel vapor inhalation exposures. Inhalation chamber fuel vapors were generated by passing the fuel through a column heated to 120°F. Three animal species were employed (mice, rats and dogs) with both sexes being represented by rats and dogs. A summary of the clinical tests performed on rats and dogs at the end of the exposure period and on rats at the end of two years is shown in the slide. Forty-two tissue samples are being taken from all of the animals at the 90-day point and from the mice and rats at the two-year point. Tissues and blood samples taken at the two-year point are intended to provide information on the long-term effects of fuel exposure and also on the tumor formation

Table III. Fuels Comparative Toxicity Study - Inhalation Protocol Summary

Laboratory:	NMRI/AFAMRL/THRU
Duration:	90-day continuous vapor inhalation
Animal Complement:	(At two concentrations per fuel)
	Rats: 75 male and 75 female (50 of each held for two years)
	Mice: 150 female
	Dogs: 3 male and 3 female
Clinical Tests:	Body weight, organ weight, hematology, clinical chemistries
Histopathology:	90 days - 42 tissues 2 years - 42 tissues

Table IV. Fuels Comparative Toxicity Study-Summary of Inhalation Studies at THRU

Fuel	Test Concentrations
Petroleum JP-5	750 mg/m^3 150 mg/m^3
Shale JP-5	750 mg/m^3 150 mg/m^3
Petroleum DFM	300 mg/m^3 50 mg/m^3
Shale DFM	300 mg/m^3 50 mg/m^3

potential. Table IV indicates the inhalation chamber fuel concentrations. There was one unexposed control group for each fuel tested.

Before I describe the results of these experiments up to this time, I think a view of the schedule and milestones is appropriate. Figure 1 depicts the time schedule we have been following. As you can see the exposure phase of all four experiments has just recently been completed. Clinical test results at the 90-day point for each fuel are complete and results of the clinical tests from the two-year, long-term follow-up of the petroleum fuels are complete. Only petroleum fuels 90-day histopathology reports are complete, however. Both Air Force and Navy demands at the Toxic Hazards Research Unit have saturated the pathology branch capability causing some delay. The target date for the four fuel final reports, including the two-year long-term effects follow-up, is late 1982. As you can see a very large part of the shale fuel work is incomplete at this time.

RESULTS

Since the shale fuel work is incomplete I have only partial results to report at this year's meeting.

Table V shows the results of the ocular and dermal irritation studies. Reaction to the ocular and dermal irritation in the case of all four fuels was mild. Mild dermal irritation observed was at the abraded test sites. Only shale DFM produced any signs of eye irritation and that occured in only one rabbit.

Table V. Fuels Comparative Toxicity Study - Skin and Eye Irritation Results

Fuel	Skin[a]	Eye
Petroleum JP-5	none	none
Shale JP-5	none	none
Petroleum DFM	none	none
Shale DFM	none	slight

[a]Intact and abraded.

Skin sensitization results are summarized in Table VI. Sensitization potential is a measure of the number of test animals exhibiting a positive reaction out of the 24 animals treated. Slight sensitization potential represents one to

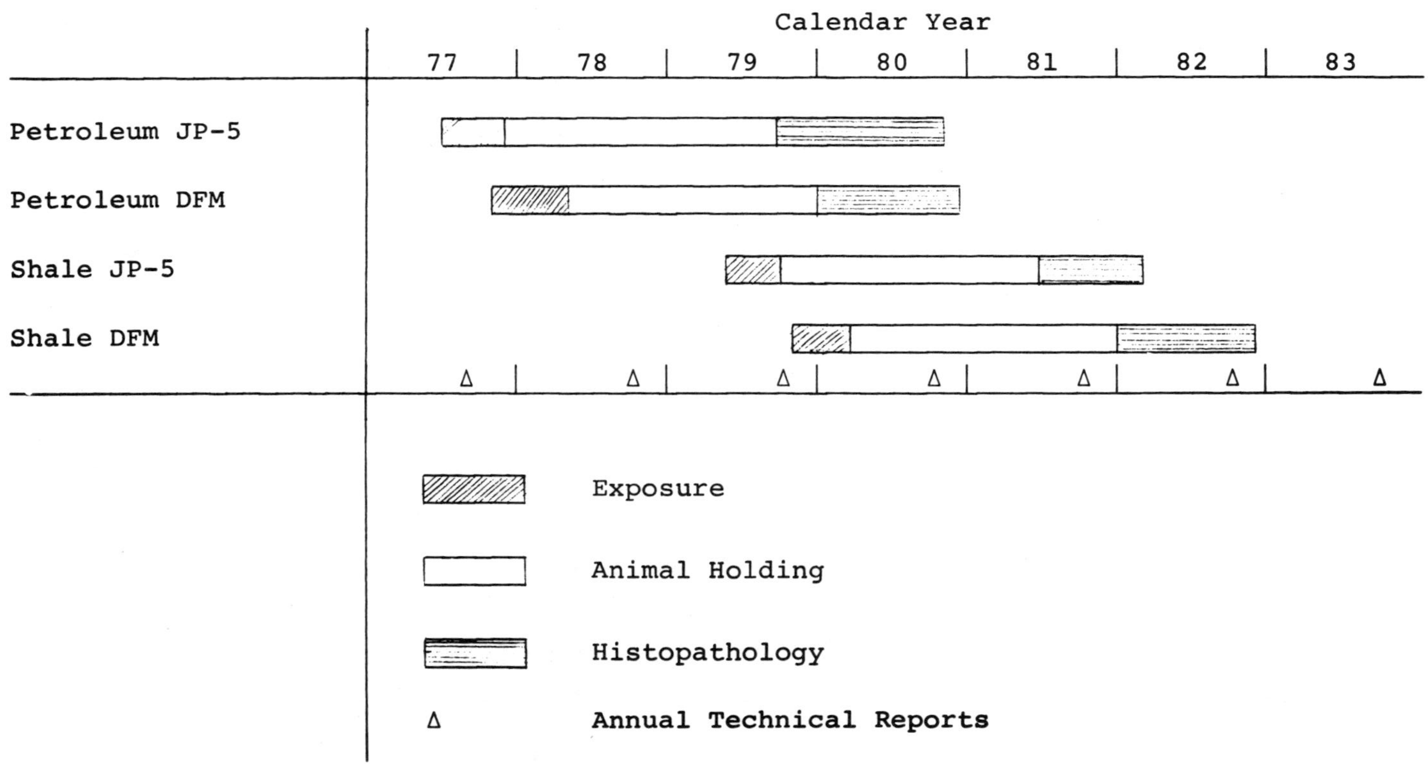

Calendar Year
77
78
79
80
81
82
83
Petroleum JP-5
Petroleum DFM
Shale JP-5
Shale DFM
Exposure
Animal Holding
Histopathology
Annual Technical Reports

three animals responding and moderate represents four to 10 animals responding. Sensitization reaction is a measure of the severity of the reaction of those animals responding. None of the fuels tested were judged to be sensitizers according to the test criteria.

The effects of fuel vapor exposure on rat weight gain are summarized in Table VII. Rats exposed to the shale fuels have not been on the study long enough to see any general health trends. The only trend we can see from this incomplete data is that male rats may be more affected by fuel vapors than female rats regardless of the fuel source. Dogs were sacrificed after the 90-day exposure and there were no apparent effects from either of the four fuels during that time.

The histopathology report of rats exposed to petroleum JP-5 and sacrificed at 90 days indicated evidence for kidney injury in both exposure groups of male rats. The injury consisted of nephropathy characterized by multifocal tubular atrophy and of focal tubular necrosis at the cortico-medullary junction. Both lesions were more severe in the 750 mg/m^3 petroleum JP-5 exposure groups and appeared to be related effects, in that the more severe the nephropathy was, the more sever the tubular necrotic lesions. These changes were not seen in unexposed controls or in female rats. Neither were seen in the dogs. Table VIII is a summary of the significant clincial measurements taken at the 90 day sacrifice. The table shows only male rat data. Statistical significance expressed here is measured between the fuel exposure group and its control and not between petroleum and shale exposure groups. The female rat data although exhibiting what appeared to be some random deviations from normal were as a whole within normal limits for the species. Male rat chemistry changes shown in this slide were considered consistent with the histopathologic findings of kidney injury.

The histopathology report of rats exposed to petroleum DFM indicated hyaline degeneration of the cytoplasm of the convoluted renal tubules in 68% of low dose male rats and 84% of high dose male rats. The lesion was absent in exposed females and all controls. This lesion was considered a mild reversible degenerative process that resulted from accumulation of protein in the cytoplasm of convoluted tubule cells due to an incapacity of the cells to transfer resorbed blood protein from glomerular filtrate back into the blood. An associated lesion (nephropathy) was seen in 96% of the high dose group male rats. High dose female rats plus all low dose and control rats were free of this change. The nephropathy was described as the presence of granular casts within

Table VI. Fuels Comparative Toxicity Study - Skin Sensitization Summary

Fuel	Sensitization[a] Potential	Sensitization[b] Reaction
Petroleum JP-5	Moderate	Mild
Shale JP-5	Slight	Mild
Petroleum DFM	Moderate	Mild
Shale DFM	Moderate	Mild

[a]A measure of the number of test animals exhibiting a positive reaction.

[b]A measure of the severity of reaction:

Score:	
0- 25	No Reaction
26- 99	Mild
100-200	Moderate
200	Severe

Table VII. Fuels Comparative Toxicity Study - Rat Body Weight Effects Summary

Fuel		Male		Female
Petroleum JP-5				
750 mg/m^3	$\downarrow$ }	recovered		---
150 mg/m^3	$\downarrow$ }			---
Shale JP-5				
750 mg/m^3	$\downarrow$ }	incomplete		---
150 mg/m^3	$\downarrow$ }			---
Petroleum DFM				
300 mg/m^3	$\downarrow$	no recovery	$\downarrow$	no recovery
50 mg/m^3	$\downarrow$	recovery	$\downarrow$	from 4 mo postexposure
Shale DFM				
300 mg/m^3	$\downarrow$ }	incomplete	$\downarrow$ }	incomplete
50 mg/m^3				

Arrows express trends seen in experimental animals compared to controls.

TABLE VIII. FUELS COMPARATIVE TOXICITY STUDY - 90-DAY MALE RAT CLINICAL MEASUREMENTS

	Petroleum JP-5			Shale JP-5		
	Control	150 mg/m^3	700 mg/m^3	Control	150 mg/m^3	700 mg/m^3
BUN	14.9	17.0[a]	18.5[a]	17.8	17.4	20.3[a]
Creatinine	0.53	0.55	0.62[a]	0.66	0.76[b]	0.78[b]
A/G Ratio	2.3	2.3[a]	2.0[a]	1.37	1.2[b]	1.2[b]

[a]Statistically different from controls at < 0.05.

[b]Statistically different from controls at < 0.01.

convoluted renal tubules in 68% of low dose male rats and 84% of high dose male rats. The lesion was absent in exposed females and all controls. This lesion was considered a mild reversible degenerative process that resulted from accumulation of protein in the cytoplasm of convoluted tubule cells due to an incapacity of the cells to transfer resorbed blood protein from glomerular filtrate back into the blood. An associated lesion (nephropathy) was seen in 96% of the high dose group male rats. High dose female rats plus all low dose and control rats wree free of this change. The nephropathy was described as the presence of granular casts within viable collecting tubules localized at the corticomedullary junction of the kidney. This change was seen in connection with a variable and mild chronic interstitial nephritis. Review of the 90 day sacrifice clinical measurements taken from rats exposed to DFM fuels did not bear out the histopathologic findings.

At the end of the two year follow-up period, ratse exposed to petroleum fuels did not exhibit increased mortality.

DISCUSSION

Because the shale fuels inhalation phase of the comparative study has only recently been completed and not all of the data is available, a comparison of shale versus petroleum fuel inhalation toxicity cannot be made at this time. The irritation studies are complete and the results indicate all four fuels could be considered nonirritants. Results of the skin sensitization studies indicate all four were nonsensitizers.

Although it may be premature to speak about the toxicity of a material that chemically may vary from batch to batch or change within a batch upon storage, it has been our belief that at the very least the work will serve as a benchmark from which to compare later shale fuels. At the very best we may hope for a clean bill of health for a much needed alternate fuel.

REFERENCES

1. U. S. Navy-ENRRDO, 1976a, U. S. Navy Energy R&D Progress, Washington, D.C.

2. Military Specification, 1976, Turbine fuel, aviation grades JP-4 and JP-5, MIL-T-5624K.

3. Military Specification, 1973, Fuel, oil, diesel, marine, MIL-F-16884G.

4. J. H. Draize, Dermal Toxicity, Appraisal for the Safety of Chemicals in Foods, Drugs, and Cosmetics. The Staff of the Divsiion of Pharmacology of the Federal Food and Drug Administration, Austin, Texas, the Editorial Committee of the Association of Food and Drug Officials of the United States, 1959, p. 51.

5. K. Landsteiner, and Jacobs, J. Studies on Sensitizations of Animals with Simple Chemical Compounds. J. Exp. Med. 61, 643-656 (1935).

CHAPTER 11

ACUTE TOXICITY OF SELECTED CRUDE AND REFINED SHALE OIL- AND PETROLEUM-DERIVED SUBSTANCES

L. H. Smith, W. M. Haschek, and H. Witschi, Biology Division, Oak Ridge National Laboratory, Oak Ridge, Tennessee 37830

INTRODUCTION

To help evaluate and predict hazards of substances to which humans may be exposed in the workplace, several kinds of tests with laboratory animals can be used. These tests are usually designed to provide information on overall toxicity of a given compound and to detect its potential to cause untoward effects.

It was the objective of the present study to obtain general information on the toxicity of selected samples of Crude Paraho Shale Oil and some of its derivatives, some crude petroleums, and three refined petroleum products. Five tests were used to determine the acute toxicity of these substances: acute lethality in mice following oral or intraperitoneal administration of a single dose; acute dermal toxicity of a single dose in rats; delayed-type allergic contact hypersensitivity in guinea pigs; primary eye irritation and primary skin irritation of a single dose in rabbits. In addition, we have examined histopathologic changes induced in mice following intraperitoneal (ip) injection of a single large dose of Crude Shale Oil and two of its hydrotreated derivatives. Studies also have been initiated to examine the tumor inducing potential of selected samples in the mouse lung adenoma bioassay. The present report describes our findings and shows that all compounds tested have very low or no acute toxic effects in laboratory animals.

---

*Research sponsored by the Office of Health and Environmental Research, U. S. Department of Energy, under under contract W-7405-eng-26 with the Union Carbide Corporation.

MATERIALS AND METHODS

Substances Evaluated

Bulk samples of the materials were received from the Chemical Repository. Viscous samples were thoroughly mixed by shaking after being warmed in a water bath at 70-80°F. Where feasible, samples were applied undiluted. For the experiments in which it was necessary to administer graded doses, measured portions were diluted with corn oil (laboratory grade, Fisher Scientific) to provide the desired final concentrations; the density of all substances was assumed to be 1.0.

The following substances were studied:

ORNL Repository No.	Name
Shale oil and its derivatives	
4601	Crude Shale Oil
4602	Hydrotreated Shale Oil
4607	Hydrotreated Residue
4608	JP-5 Product (jet fuel)
4609	JP-8 Product (jet fuel)
4610	Diesel Fuel Marine Product
Petroleum and its derivatives	
5107	Mixed Petroleum Crudes
5301	Wilmington Crude Oil
5305	Recluse Crude Oil
4614	JP-5 Product (jet fuel)
4615	JP-8 Product (jet fuel)
4616	Diesel Fuel Marine Product

Acute LD_{50} in Mice

The objective of this test was to determine acute toxicity following a single oral or ip administration of a substance. The animals used were BALB/c male or female mice that weighed 20-25 g.

Mice were randomly grouped six per cage and were fasted for 18 hr before administration of a substance. For oral administration a ball-tipped needle (20 gauge) fitted to a 1-ml syringe was used. The administration volume was 10 μl/g body weight except when a dose of 16 g/kg was delivered; in this case 20 μl/g body weight of a solution containing 0.8 ml/ml of the test agent was administered. The doses were 2,

4, 8, and 16 g/kg. For ip injections, a 1-ml syringe fitted with a 22-gauge needle was used. The injection volume was 10 l/g body weight, and doses were 1, 2, 4, and 8 g/kg. Additional doses were tested when necessary. Corn oil was the vehicle used for both routes of administration.

Animals were returned to their cages with free access to food and water and were observed for signs of toxicity. The number of survivors was recorded daily. Two weeks after administration of the substances, the surviving mice were killed. The LD_{50} was calculated from the number of 14-day survivors according to the procedures of Weil.[1]

Acute Dermal Toxicity in Rats*

The objective was to determine the acute toxicity of a substance following a single application to the skin. The animals used for this test were albino male (300-400 g) and female (200-250 g) rats. Each substance was tested on five or six animals of each sex.

The animals were lightly anesthetized with ether, and the hair was clipped from the sides and back. The substance was applied to the center of the back and spread over an area of 25-50 cm^2. An attempt was made to cover the same area on each animal and to apply the substance uniformly. A microliter pipetting device was used to apply 2 ml/kg body weight of undiluted sample. The rats were then placed in individual cages where they remained for 4 hr. During this time they were observed for signs of systemic toxicity. At the end of the 4 hr period any residual substance was removed and signs of skin injury were noted. The animals were returned to maintenance cages and observed over a period of 14 days. At the end of this period, body weights were obtained and necropsies were performed.

Delayed-Type Allergic Contact Hypersensitivity in Guinea Pigs*

The objective was to determine the delayed-type allergic response resulting from cutaneous contact with a substance. The animals used for this test were Hartley albino male and female guinea pigs about 3 months old. Each substance was tested on four different animals of each sex.

Hair was clipped from a 3- to 4-cm^2 area on the back between the shoulder blades. Clipping was done on the first day of injection and when necessary during the course of further injections. A 0.1-ml portion of a 0.1% solution of the substance in 20% DMSO-80% saline or in propylene glycol was injected intradermally three times a week for 3 weeks. The injections were within a 3- to 4-cm^2 field at different sites within the field. Twelve days after the ninth injection, a challenge dose of 0.05 ml of the 0.1% solution of the substance was injected intradermally into a fresh site of the clipped area. Readings for skin reactions were made 24 and 48 hr after each of the nine injections and after the challenge dose. The response was graded and evaluated according to the method of Draize et al.[2] which scores sensitization in terms of redness, edema, blistering, and necrosis.

Primary Eye Irritation in Rabbits*

The objective was to determine surface irritation to ocular tissues by a substance following a single application. The animals used were New Zealand albino male rabbits that weighed about 2 kg. Each substance was tested on one eye of four animals; the other eye served as a control. The lower lid was gently pulled away from the eye, and 0.1 ml of the undiluted agent was instilled into the conjunctival sac by use of a pipetting device. The eyes were examined for signs of irritation 1, 2, and 3 days after instillation. The response was graded and evaluated according to a modified Draize method[2] which scores irritation in terms of redness of the lid and the conjunctiva, edema of the lid and conjunctiva, and opacity of the eye surface.

Primary Skin Irritation in Rabbits*

The objective was to determine skin irritation by a substance following a single application. The animals used for this test were New Zealand albino male rabbits that weighed about 2 kg. Each substance was tested on six different animals.

Hair was clipped from the sides of the rabbits, and 0.5 ml of the undiluted substance was distributed evenly over a 6- to 7-cm^2 area of flank skin by use of a pipetting device. The area was covered with a gauze pad secured with strips of adhesive tape, and the flank was wrapped with an elastic bandage. Four hours later, the bandage and pad were removed and any residual substance was wiped off with dry gauze. At this time the first reading for a skin response was made, followed by further readings at 24 and 72 hr after application. If a definite response was observed at 72 hr, another reading was made 7 days after application. The response was graded and evaluated according to the method of Draize et al.[2] which scores irritation in terms of redness and edema of the skin.

Histopathology in Mice

Male BALB/c mice 6-8 weeks old, 10 per group, were given a single ip injection of the test agent. The mice were killed 2 or 4 days later, and gross necropsy was performed. Tissues were removed and fixed in 10% buffered formalin. They were embedded in paraffin, sectioned at 4-6 μm, and stained with hematoxylin and eosin. The following organs were examined histologically: brain, heart, liver, kidney, esophagus, stomach, intestine, colon, pancreas, spleen, gall bladder, adrenal, lung, trachea, lymph node, and thymus.

Lung Adenoma Assay in Strain A Mice

The objective was to determine if multiple ip injections of a substance can induce lung adenomas in strain A/Jax male mice 6-8 weeks old. In preliminary toxicity studies, the maximum tolerated dose (MTD) was determined. The MTD is defined as the highest dose producing no mortality when given in a total of six injections over a 2-week period. After 6 weeks, survivors were killed and gross necropsies were performed. In the adenoma test proper, each substance was injected into 30 mice three times a week (Monday, Wednesday, Friday) for 8 weeks. The following doses were used: the MTD, one-half the MTD, and one-fifth the MTD. Control groups were a) 30 mice injected with the same vehicle (corn oil) used to dissolve the test substance, b) 50 mice given a single injection of urethan at 1 g/kg body weight (positive control), and c) 50 mice which received nothing.

Body weights were recorded every 2 weeks, and the mice were killed 20 weeks after the first injection. The lungs were fixed in Tellyesniczky's solution and the number of tumors on the lung surface was counted. Selected tissues were fixed in 10% buffered formalin and Bouin's fixatives for histopathology.

RESULTS

Acute Toxicity in Mice and Rats

Data for acute toxicity of shale oil and its derivatives are given in Table I. Crude Shale Oil was the only one that produced acute deaths in both male (at doses of 8 and 10 g/kg) and female (at doses of 4,8, and 16 g/kg) mice when given orally. Following ip injection, deaths also occurred at doses of 2 g/kg. Because the mortality data were similar for both sexes, substances 4608-4610 were tested in female mice only. The LD_{50} was between 6 and 8 g/kg for substances 4608 and 4609 and greater than 16 g/kg for substance 4610.

Table I. Acute Lethality Test in Mice: Shale Oil and Its Derivatives

Substance	LD_{50} (g/kg) Oral		Intraperitoneal	
	Male	Female	Male	Female
4601	11.3	11.3	6.1 (4.5-8.2)[a]	4.3 (2.0-6.1)
4602	>16.0	>16.0	>16.0	>16.0
4607	>16.0	>16.0	>16.0	>16.0
4608	ND[b]	>16.0	ND	8.0 (5.4-12.8)
4609	ND	>16.0	ND	6.4 (5.0-8.0)
4610	ND	>16.0	ND	>16.0

[a]95% confidence limits.

[b]ND = not done.

Results for petroleums and derivatives are shown in Table II. Acute toxicity data were for all practical purposes similar to the ones observed with shale oil and its derivatives.

In both male and female rats, there were no visible toxic responses arising from the dermal application of 2 g/kg of samples 4601, 4602 and 4607. Neither visible skin lesions nor manifestations of central nervous system pertubations were observed. Deaths did not occur during the 2 weeks following application of the substances, and all rats showed a progressive gain in body weight (Table III). Similar observations were made when rats were treated with 2 g/kg of Crude Shale Oil- or petroleum-derived JP-5, JP-8, or Diesel Fuel Marine product: no animals died, no skin lesions developed, and weight gains over a 2-week period were normal.

Delayed-Type Contact Hypersensitivity

Guinea pigs were sensitized with Crude Shale Oil, Hydrotreated Shale Oil, Hydrotreated Residue, or shale oil- and petroleum-derived jet and diesel fuels. The data are summarized in Table IV. During the sensitization procedure, only slight redness and focal scab formation were noted in some of the animals for all nine substances; edema and necrosis were not seen. After the challenge injection, all animals sensitized with Shale Oil, Hydrotreated Shale Oil, and Hydrotreated Residue developed a degree of redness and edema greater than that observed after the sensitizing injections. The challenge injection of these three substances also produced a slight necrosis in most guinea pigs, and for two of the substances, a marked lesion with a necrotic center developed. No difference in reaction between the sensitizing and challenge injection was observed in animals exposed to the shale oil- or petroleum-derived jet and diesel fuels.

Eye and Skin Irritation in Rabbits

The data in Table V show that only the shale oil and to a lesser extent its hydrotreated derivative produced a visible redness of the rabbit eye. However, irritation was slight and transient and is considered insignificant in evaluating substance toxicity. None of the other substances produced visible irritation at the three observation intervals. There were no indications that the shale oil, its five derivatives, or the petroleum products were irritating to rabbit skin to the extent of producing visible redness or edema (Table VI).

Table II. Acute Lethality Test in Mice: Petroleums and Their Derivatives

Substance	LD_{50} (g/kg)			
	Oral		Intraperitoneal	
	Male	Female	Male	Female
5107	>10.0	ND[a]	ND	ND
5301	>16.0	ND	ND	ND
5305	>16.0	ND	ND	ND
4614	ND	>16.0	ND	11.2 (8.13-15.5)[b]
4615	ND	>16.0	ND	8.0 (5.5-12.8)
4616	ND	>16.0	ND	>16.0

[a]ND = not done.

[b]95% confidence limits.

Table III. Weight Gain in Male and Female Rats Following Dermal Application of 2 g/kg of Shale Oil, Hydrotreated Shale Oil, or Hydrotreated Residue[a]

Substance	Mean Body Weight (g) ± SE			
	Male		Female	
	Day 0	Day 14	Day 0	Day 14
4601	358 ± 8	415 ± 7	209 ± 5	241 ± 2
4602	352 ± 9	422 ± 9	209 ± 3	238 ± 4
4607	370 ± 4	435 ± 7	225 ± 5	256 ± 6

[a]Six animals per group except for substance 4602 females, for which there were 5 rats.

Table IV. Delayed-Type Allergic Contact Hypersensitivity Test in Guinea Pigs

Substance	Average response[a] 24 and 48 hr after each of nine sensitizing injections			Average response[a] 24 and 48 hr after the challenge injection		
	Redness	Edema	Necrosis	Redness	Edema	Necrosis
Shale oil and its derivatives						
4601	±	0	0	1-2	2	±
4602	0	0	0	2	1	±
4607	0	0	0	2	1	±
4608	0-1	0	0	0	0	0
4609	0-1	0	0	0	0	0
4610	0-1	0	0	0	0	0
Petroleum derivatives						
4614	0-1	0	0	0	0	0
4615	0-1	0	0	0	0	0
4616	0-1	0	0	0	0	0
Propylene glycol[b]	0-1	0	0	0	0	0
Dinitrochlorobenzene[c]	1-2	0-1	1-2	1-2	1-2	2-3

[a]Scale: 0, ±, 1-4.
[b]Vehicle.
[c]Positive control.

Table V. Primary Eye Irritation Test in Rabbits

Substance	Response at Indicated Times after Instillation[a]		
	24 hr	48 hr	72 hr
Shale oil and its derivatives			
4601	6 of 6 showed grade 1 redness[b]	0	0
4602	2 of 6 showed grade 1 redness[b]	0	0
4607	0	0	0
4608	0	0	0
4609	0	0	0
4610	0	0	0
Petroleum derivatives			
4614	0	0	0
4615	0	0	0
4616	0	0	0

[a]Four or six eyes per substance.

[b]On the basis of 0-4 response of lid, conjunctiva, and cornea.

Table VI. Primary Skin Irritation Test in Rabbits

Substance	No. of Rabbits Responding at Indicated Time After Application[a]					
	4 hr		24 hr		72 hr	
	Redness	Edema	Redness	Edema	Redness	Edema
Shale oil and its derivatives						
4601	0	0	0	0	0	0
4602	0	0	0	0	0	0
4607	0	0	0	0	0	0
4608	0	0	0	0	0	0
4609	0	0	0	0	0	0
4610	0	0	0	0	0	0
Petroleum derivatives						
4614	0	0	0	0	0	0
4615	0	0	0	0	0	0
4616	0	0	0	0	0	0

[a]Six animals per substance.

Histopathology

Mice killed 2 or 4 days following ip injection of 4601 at 4 g/kg, or 4602 or 4607 at 16 g/kg showed superficial pyogranulomatous serositis of the peritoneal cavity resulting from local reaction to the substances. Lymphoid depletion of the thymic cortex was present in all animals at 2 and 4 days, but was most marked in mice injected with 4602. Serositis and thymic cortical depletion were the only histopathologic alterations observed for these substances. No lesions were present in animals injected with corn oil alone.

Lung Adenoma Assay

This assay was done essentially as described by Shimkin and Stoner.[3] In preliminary toxicity studies, the MTD for 4601 was determined to be 2.5 g/kg per injection and for 4602, 16 g/kg. However, during the course of the adenoma assay it was found that the MTD, as determined, was too high. Only in one group was the survival rate higher than 50%. Cumultive mortality rates are shown in Figure 1A and B, and the average weight gain of survivors is shown in Figure 2. It is interesting to note that with both Crude Shale Oil and Hydrotreated Shale Oil animals continued to die even after cessation of exposure to the test substances. The reasons for this delayed toxicity remains to be established. Treated animals also generally failed to gain weight at rates comparable to that of controls.

As shown in data of Table VII, tumor incidence was 30-60% in treated animals and not significantly different from animals injected with vehicle (corn oil) alone. On the other hand, tumor multiplicity was significantly higher in animals injected with 1.25 g/kg of Crude Shale Oil. In all other groups, the differences from the control value were statistically not significant. However, few animals survived when injected with 2.5 g/kg of Crude Shale Oil or with 16, 8, or 3.2 g/kg of Hydrotreated Shale Oil. In view of this general toxicity, the adenoma assay in these groups is probably not representative and needs to be repeated at lower dose levels. Detailed results on histopathology will be reported later.

DISCUSSION

Our data show that, when ingested, Crude Shale Oil may be considered slightly toxic and all other samples are practically nontoxic; no sex difference was found where both male and female mice were tested. Following ip injection the substances were slightly toxic. Acute skin toxicity was invariably greater than 2 g/kg, and no evidence was obtained to document appreciable eye and skin irritation. Histopatho-

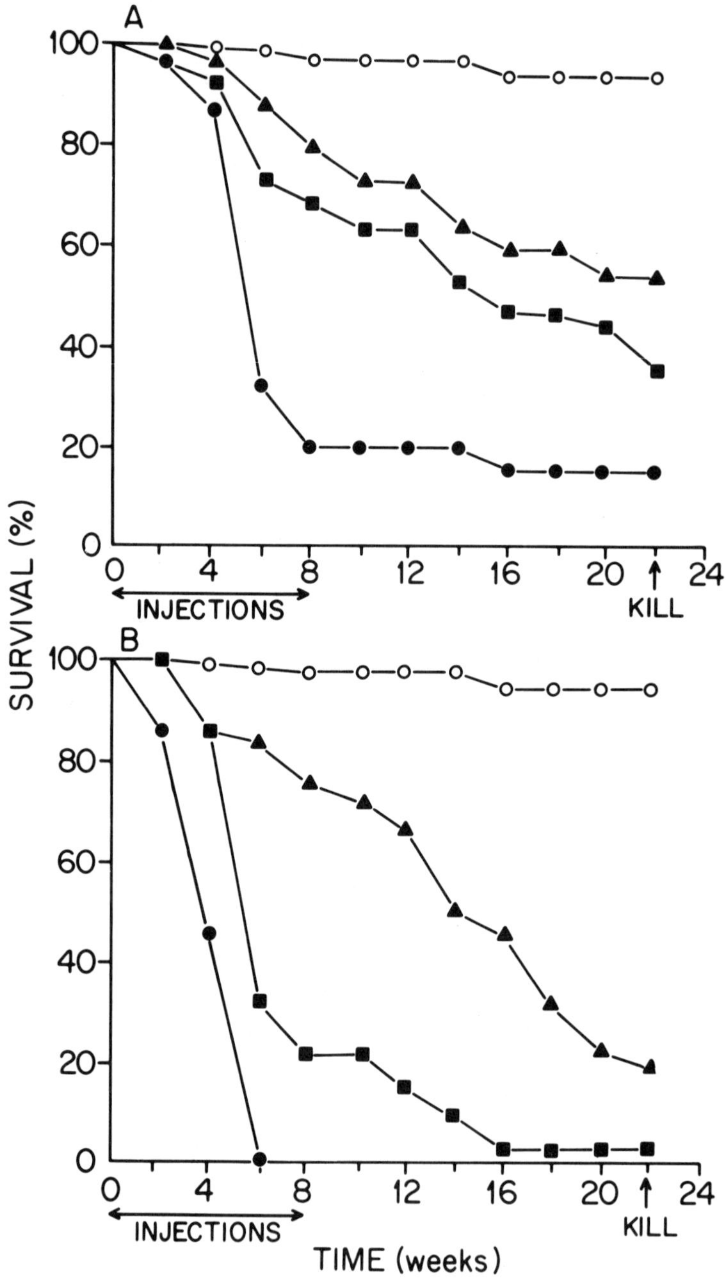

Figure 1. Survival of Mice Given (A) 24 Injections of Crude Shale Oil 4601 or (B) Hydrotreated Shale Oil 4602. (A): ●, 2.5 g/kg; ■, 1.25 g/kg; ▲, 0.5 g/kg; ○, Corn Oil. (B): ●, 16 g/kg; ■, 8 g/kg; ▲, 3.2 g/kg; ○, Corn Oil. Brackets Indicate SE's.

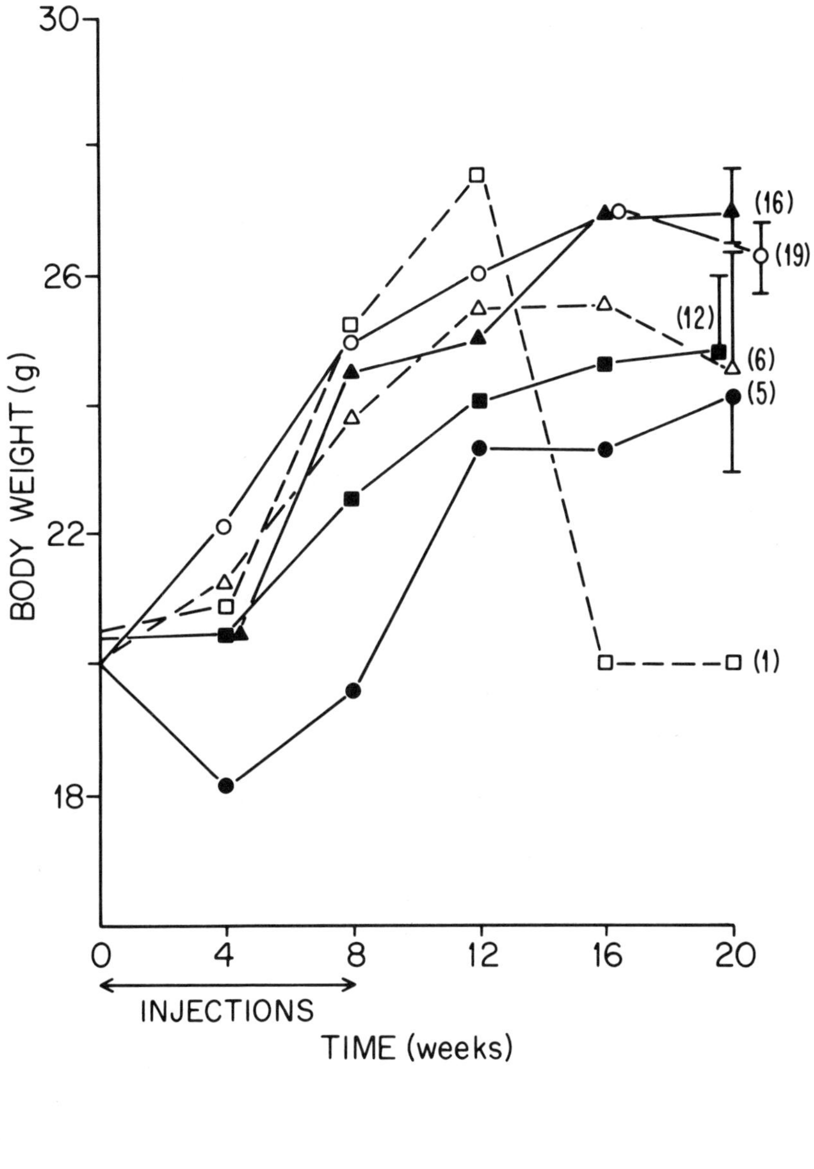

Figure 2. Body Weight of Mice Given 24 Injections of Crude Shale Oil 4601 or Hydrotreated Shale Oil 4602. 4601: ●, 2.5 g/kg; ■, 1.25 g/kg; ▲, 0.5 g/kg. 4602: □, 8 g/kg; △, 3.2 g/kg; ○, Corn Oil. Number of Mice Shown in Parentheses. Brackets Indicate, SE's.

Table VII. Mouse Lung Adenoma Assay: Shale Oil and Hydrotreated Shale Oil

Substance	Dose per Injection[a] (g/kg)	No. of Mice Initially	No. of Survivors	No. of Mice With Tumors	No. of Tumors/Mouse[b]
4601	2.5	30	5	3	0.6 ± 0.3
	1.25	30	12	6	1.3 ± 0.5
	0.5	30	16	7	0.6 ± 0.2
4602	16	30	0	-	-
	8	30	1	0	0
	3	30	6	3	1.0 ± 0.6
Corn oil	10	20	19	6	0.4 ± 0.1
Untreated	-	20	20	3	0.2 ± 0.1
Urethan[c]	1.0	20	20	20	23.2 ± 1.7[d]

[a]Given three times a week for 8 weeks.
[b]Total number of tumors divided by total number of surviving animals; mean $\pm$ SE.
[c]One ip injection of 1 mg/g body weight.
[d]$p < 0.05$ compared with corn oil control.

logic findings following acute administration of large amounts of test material were not remarkable. Lympoid reduction in the thymic cortex was interpreted as a stress reaction rather than a substance-specific effect.

Compared with petroleum and some of its derivatives, shale oil and its derivatives appear to have practically the same overall low toxicity. Fuels from both sources have about the same LD_{50}'s and are essentially nonirritating to the skin and eye. These observations corroborate our earlier studies which showed that three other petroleum-based substances did not elicit skin or eye toxicity (Table VIII). We would not expect, therefore, that acute exposure to these jet and diesel fuels produced from shale oil would be any more hazardous than exposure to their petroleum-derived equivalents currently in use.

Table VIII. Skin and Eye Toxicity of Petroleum-Based Substances

Substance	Chemical Repository No.	Acute LD_{50}, Skin (g/kg)	Eye Irritation	Skin Irritation
Syncrude	1106	> 2	0	0
Mixed Petroleum Crude	5107	> 2	0	0
Synthoil Crude	1201	> 2	0	0

From the present data, we conclude that a single exposure of humans to any one of the six Paraho/SOHIO substances tested would result in little if any systemic toxicity or skin or eye injury. One possible exception arises from the findings of the skin-sensitization bioassay. Our data suggest that repeated topical exposures to the Crude Shale Oil or the hydrotreated derivatives 4602 or 4607 could possibly elicit an immunologically mediated reaction in humans. This type of reaction which could take weeks or months to develop may become manifest as itching, reddening, eruption, and edema of the skin. These reactions may disappear if there is no further contact with the sensitizing substance, but the state of sensitization may be permanent.

Table IX. Acute Oral Lethality Test of Selected Coal Conversion Products in BALB/c Male Mice [from Ref. 5]

Substance	Chemical Repository No.	LD_{50} (g/kg)
H-Coal		
H-coal Distillate (raw)	1601	3.6 (2.8-4.5)[a]
HDT Distillate (low severity)	1602	4.0 (3.4-4.7)
HDT Distillate (high severity)	1604	5.5 (3.8-7.2)
HDT Fuel Oil (low severity)	1617	6.2 (3.7-8.7)
HDT Fuel Oil (high severity)	1619	6.2 (5.1-7.3)
Atmospheric Still Bottoms (ASB) (Fuel Oil Mode)	1313	2.3 (1.9-2.6)
Vacuum Still Overhead (VSOH) (Fuel Oil Mode)	1314	2.6 (2.2-3.2)
Atmospheric Still Bottoms (ASB) (Syncrude Mode)	1309	3.6 (2.4-5.2)
Vacuum Still Overhead (VSOH) (Syncrude Mode)	1310	2.5 (1.7-3.1)
Coal SRC I		
HDT Light Organic Liquid (low severity)	1606	3.6 (3.1-4.1)
HDT Light Organic Liquid (high severity)	1608	10.0
HDT Recycle Solvent (low severity)	1614	2.6 (1.9-3.2)
HDT Recycle Solvent (high severity)	1616	5.3 (4.4-6.3)
Coal SRC II		
Fuel Oil Blend	1701	3.1 (2.7-3.5)
Zinc halide Hydrocracking Process		
Product Distillate	1801	3.4 (0.5-2.3)

[a]95% confidence limits

The data obtained in the lung adenoma bioassay suggest that Crude Shale Oil has tumorigenic potential. This appears to agree with preliminry data obtained in skin painting experiments.[6] Before any more definitive conclusions can be drawn, particularly as to the carcinogenic potency of shale oil, additional studies with the lung adenoma bioassay will have to be completed.

It is interesting to note that the oral LD_{50}'s of shale oil, petroleum, and their derivatives are consistently higher than the oral LD_{50} found with materials derived from various coal conversion processes. Table IX lists some data for comparative purposes.

In conclusion, acute toxicity of Paraho Crude Shale Oil and of five derivatives does not appear to be a problem of immediate concern. The substances produced slight acute effects similar to those in a report by Weaver and Gibson, except that they failed to obtain a positive response in the assay for skin sensitization. This discrepancy might need to be resolved in further studies.

REFERENCES

1. C. S. Weil, "Tables for Convenient Calculation of Median-Effective Dose (LD_{50} or ED_{50}) and Instructions in Their Use," Biometrics, 8, 249-263 (1952).

2. J. H. Draize, G. Woodward, and H. O. Calvery, "Methods for the Study of Irritation and Toxicity of Substances Applied Topically to the Skin and Mucous Membranes," J. Pharmacol. Exp. Ther., 82, 377-390 (1944).

3. M. B. Shimkin and G. D. Stoner, "Lung Tumors in Mice: Application to Carcinogenesis Bioassay," Adv. Cancer Res., 21, 1-58 (1975).

4. N. K. Weaver and R. L. Gibson, "The US Oil Shale Industry: A Health Perspective," Am. Ind. Hyg. Assoc. J., 40, 460-467 (1979).

---

*The test methods used are based on those described in Principles and Procedures for Evaluating the Toxicity of Household Substances, National Academy of Sciences, Washington, DC, 1977. Acute toxicity, pp. 10-17; dermal toxicity, pp. 23-26; delayed-type sensitivity, pp. 36-39; eye irritation, pp. 123-124; skin irritation, pp. 28-31.

5. K. E. Cowser, et al. (ed.), "Briefing on Oak Ridge National Laboratory Life Sciences Synthetic Fuels Program," September 27, 1979, pp. 86-94.

6. J. M. Holland, These proceedings.

ACKNOWLEDGEMENT

We thank Marsha Harden, Carol Paton, and Clarissa Shelton for their excellent assistance. This research was sponsored by the Office of Health and Environmental Research, U. S. Department of Energy, under contract W-7405-eng-26 with the Union Carbide Corporation.

CHAPTER 12

SHORT-TERM MICROBIAL TESTING OF SHALE OIL MATERIALS*

T. K. Rao[a], J. L. Epler[a], M. R. Guerin[b], and B. R. Clark[b], Biology Division[a] and Analytical Chemistry Division[b], Oak Ridge National Laboratory, Oak Ridge, Tennessee 37830

INTRODUCTION

The developing oil shale industry is concerned with evaluating both short-term as well as long term health effects associated with the technology. This task includes testing of primary effluents and products in various biological test systems from which risk posed to the shale oil industrial worker as well as consumers can be assessed. The Biology and Analytical Chemistry Divisions of Oak Ridge National Laboratory have been involved in such assessment studies with a suite of samples obtained from Paraho/Sohio/-U.S. Navy. Table I illustrates various test samples and proposed biological studies with these samples. The crude oil samples, the hydrotreated oil sample along with the fuel oils are subjected to extensive analysis in a variety of animal tests (mouse), genetic tests using microbes and mammalian cells in culture, cytotoxicity and embryotoxicity tests and eventual correlation to chemical constituents determined by chemical analysis. Other test materials are to be analyzed in the microbial mutagenesis assays that are rapid and cost-effective. Confirmatory studies are suggested to be performed using the dermal carcinogenesis studies followed by chemical characterization.

Preliminary results obtained in microbial mutagenesis studies with this suite of samples are presented here.

---

*Research sponsored jointly by the Environmental Protection Agency (IAG-D5-E681; Interagency Agreement 40-516-75) and the Office of Health and Environmental Research, U. S. Department of Energy under contract W-7504-eng-26 with the Union Carbide Corporation.

Table I. <u>ORNL INVOLVEMENT</u> - Toxicological Assessment- Paraho/Sohio Refined Shale Oil

Sample Type (Repository No.)	Health Effects Bioassay												
	Acute oral	Acute dermal	Skin irritation	Eye irritation	Skin sensitization	Carcinogenesis dermal	Mutagenesis microbial	Mutagenesis *in vitro* mammalian	Mutagenesis cytogenetics	Mutagenesis mouse	Teratogenesis	Cytotoxicity-embryotoxicity	Chemical characterization bioassay preparation
Crude shale oil (4601)	X	X	X	X	X	X	X	X	X	?	X	X	X
Hydrotreated shale oil (4602)	X	X	X	X	X	X	X	X	X	?	X	X	X
Gas feedstock (4603)								X					X
JP-5 precursor (4604)								X					X
JP-5 shale oil (4608)							X	X					X
JP-5 petroleum (4614)							X	X					X
JP-8 precursor (4605)								X					X
JP-8 shale oil (4609)							X	X					X
JP-8 petroleum (4615)							X	X					X
DFM precursor (4606)								X					X
DFM shale oil (4610)							X	X	X				X
DFM petroleum (4616)							X	X					X
Acid sludge (4612)								X					X
Hydrotreated residue (4607)	X	X	X	X	X	X	X	X			X	X	X

Microbial Mutagenesis Assays of Shale Oil

The histidine reversion technique in *Salmonella typhimurium* developed by Ames et al.[1] and histidine reversion and canavanine resistance (forward mutation) in *Saccharomyces cerevisiae*[2] were used for mutagenicity determination. Metabolic activation was accomplished in these assays by incorporating mammalian (Aroclor induced rat liver) microsomal enzymes[1] into the assay.

Bacterial Mutagenesis Assays

The strain TA98 (*his* D 3052, *uvr* B, *rfa*, *pkm* 101: obtained through the courtesy of Dr. B. N. Ames)[3] designed to detect frameshift mutations was used. The test material was dissolved in sterile dimethylsulfoxide and assayed for mutagenicity in a non-toxic dose range using the pour-plate incorporation method. The assay is quantitated with respect to dose and is represented as specific mutagenic activity (rev/mg).

An "in-situ" retorted shale oil sample was fractionated using the scheme described by Swain et al.[4] as modified by Bell et al.[5] and has been previously described.[6] The neutral fraction has been subfractionated using a Florisil column. Reagent grade solvents and glass distilled water were used for extraction and chromatography.

Mutagenicity assays were performed with each primary fraction (acid, base, neutral) and the subfractionated neutral fractions. Specific mutagenic activities (rev/mg), weight distribution and weighted mutagenic activities (product of weight percent and specific activities) are given in Table II. The neutral fraction and the basic fraction contained most of the mutagenic activities.

Due to the fact that the Swain-Stedman fractionation scheme involves extreme conditions of chemical treatment, Jones et al.[7] have reported the use of liquid chromatographic separation using LH-20 Sephadex gel partitioning technique. Mutagenicity results obtained and the fractionation procedure are given in Figure 1. A total of 300 rev/mg of the starting material was obtained by the summation of various fractions. The original material gave 233 rev/mg indicating a good biological recovery. Similar observation was also made for chemical recovery.

Since the basic subfraction contained significant mutagenic activity, the subfractionation technique developed by Guerin et al.[8] was used for the isolation of mutagenic agents in the ether soluble base fraction. Figure 2 illustrates the

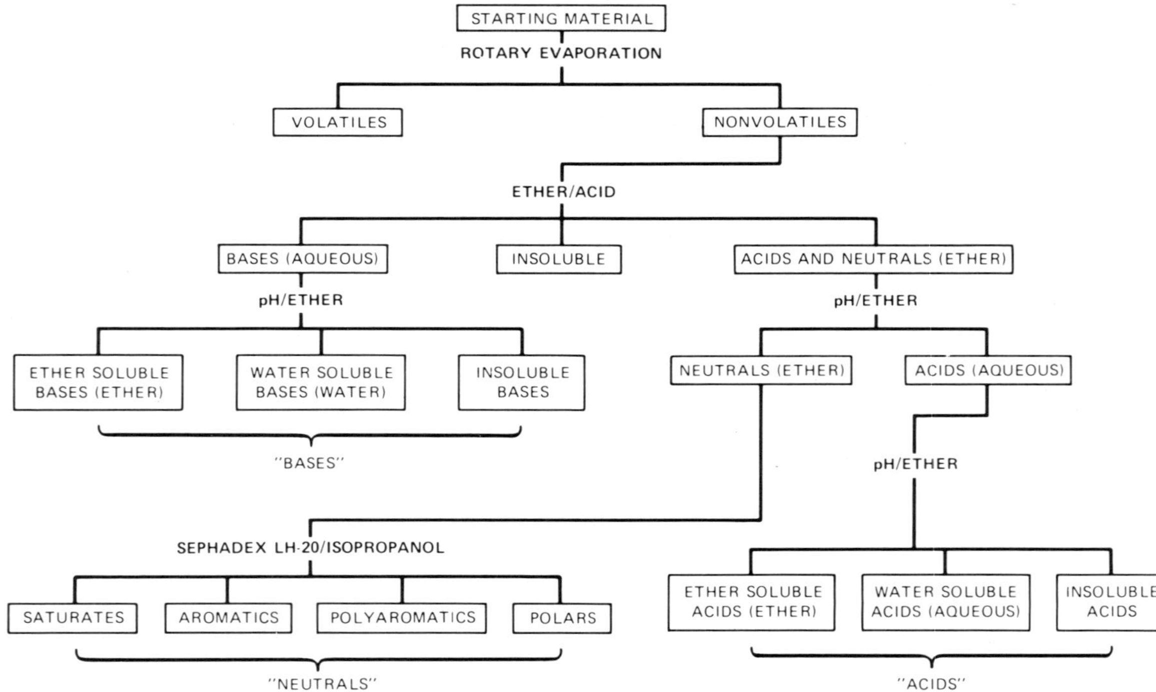

CHEMICAL CLASS FRACTIONATION FOR BIOTESTING
MATERIAL, **STEP**, (PHASE)

Figure 1. Sephadex LH-20 fractionation of shale oil coupled with mutagenicity testing. Specific mutagneic activities given were determined for the linear portions of dose response curves in the non-toxic dose range.

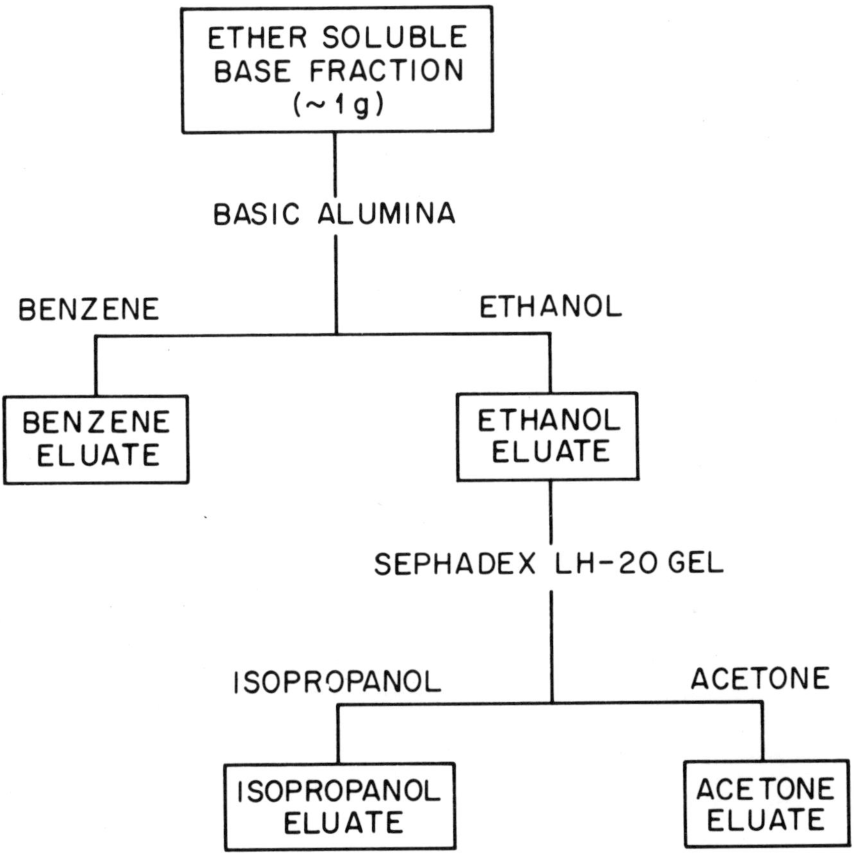

Figure 2. Subfractionation of ether soluble base fraction of shale oil using alumina and Sephadex LH-20 column chromatographics.

Table II. Distribution of Mutagenic Activity in Fractions of Simulated *in-situ* Shale Oil

No.	Fraction[a]	Relative Weight % of Total	Specific Activity (Rev/mg)[b]	Weighted Activity (Rev/mg)
1	WA_I Weak $Acids_I$	0.05	185	>1
2	WA_E Weak $Acids_E$	1.23	52	1
3	SA_I Strong $Acids_I$	0.09	0	-
4	SA_E Strong $Acids_E$	0.26	159	>1
5	SA_W Strong $Acids_W$	0.55	160	1
6	B_{Ia} $Base_{Ia}$	0.20	1377	3
7	B_{Ib} $Base_{Ib}$	0.26	800	2
8	B_E $Base_E$	7.11	952	68
9	B_W $Base_W$	0.28	223	1
	Neutral	86.66	112 (109)[c]	97
	Total	97.71		178

Neutral subfractions

No.	Fraction	Relative Weight % of Total	Specific Activity (Rev/mg)	Weighted Activity (Rev/mg)
10	Hexane			
	A	58.69	40	23
	B	2.14	625	13
	C	1.27	750	10
11	Hexane/benzene			
	A	4.38	238	10
	B	1.89	340	6
	C	1.39	320	4
12	Benzene/ether			
	A	12.43	65	8
	B	2.19	142	3
	C	1.29	253	3
13	Methanol			
	A	15.12	179	27
	B	0.49	684	3
	C	0.93	263	2
	Subtotal	102.21		112

[a] I = insoluble (fractions a and b), E = ether-soluble, W = water soluble

[b] rev/mg = revertants/milligram, the number of histidine revertants from Salmonella strain TA98 by use of the plate assay with 2 x 10^8 bacteria per plate. Values are derived from the slope of the induction curve (with Aroclor induced S-9).

[c] Activity based on assay of the total neutral fraction before chromatography rather than on the summation of the individual subfraction.

subfractionation of the ether soluble base fraction on a basic alumina column. The material is eluted with benzene followed by ethanol. Further isolation is achieved by eluting the ethanol eluate with isopropanol followed by acetone on a Sephadex LH-20 gel column. Table 3 illustrates the results obtained with an ether soluble base fraction from shale oil which show isolation of 90% of the mutagenic activity in the acetone fraction that accounts for 5% of the total weight of the starting fraction. Chemical analysis followed by mutagenesis assays have indicated aromatic amines[7] as the primary mutagenic constituents along with azaarenes and other nitrogen containing polyaromatic hydrocarbons.[9]

Table III. Subfractionation of Ether-Soluble Basic Fraction[a] from Simulated *In-Situ* Shale Oil

	Shale Oil	
	Wt. %	Rev/mg[a]
Basic fraction	100	2,500
Alumina		
Benzene	78	600
Ethanol	-	-
LH-20		
Isopropanol	12	0
Acetone	10	20,000

[a]Assayed with strain TA98 with Aroclor-induced preparation.

The Paraho Crude Shale Oil sample obtained from the Chemical Repository was subjected to chemical fractionation using the acid/base separation technique and the neutral fraction was subfractionated with Sephadex LH-20 gel chromatography into aliphatic and aromatic fractions of various ring sizes. The results obtained are given in Table IV. The base fraction and the neutral fraction were the major contributors of the mutagenic activity. By the summation of mutagenic activities of subfractions a value of 382 rev/mg of the original starting material was obtained. A suite of samples (ORNL repository numbers 4602-4616) from Paraho/Sohio refinery were tested (both raw samples and subfractions) for mutagenicity and the results are given in Table V. The hydrotreated shale oil and its residue from fractionation were not mutagenic in the bioassay. The refined oil samples such as JP-5, JP-8 and DFM product as well as their precursors were not mutagenic in the Ames assays. Two petroleum equivalents from JP-5 and DFM failed to show mutagenic activities.

Yeast Mutagenesis Assays

The results obtained with Salmonella were verified by using a multi-purpose stock of *Saccharomyces cerevisiae* (XL7-10B, α, *CAN* 1, *his* 1-7, *lys* 1-1, *ura* 1). Forward mutants for canavanine resistance (CAN → can) and revertants to histidine prototrophy were selected after treatment with the Crude Shale Oil and Hydrotreated Shale Oil fractions. The results are given in Table VI. The Crude Shale Oil was not mutagenic but the ether soluble basic fraction contained significant amount of mutagenic activity. Both the canavanine marker and the histidine marker were equally affected and in Table VI results obtained with canavanine are represented. The hydrotreated sample failed to show any mutagenic activity in either the raw sample or the ether soluble base fraction.

Comparison with Other Synfuels and Petroleum Crudes

Table VII quantitatively compares the mutagenesis results obtained with shale, coal, and petroleum derived oils in the Salmonella assay. Natural petroleum crudes are the least mutagenic with all the activity remaining in the neutral fractions. A coal derived fuel oil (SRC-II) was highly mutagenic, with the neutral fraction being the major contributor. Another oil sample, a coal hydroliquefaction product oil (H-Coal) was also mutagenic, again most of the activity remaining in the neutral fraction. Hydrotreatment of this oil has eliminated the mutagenic activity, similar to hydrotreated shale oil. Both the Paraho/Sohio Crude Shale Oil and Laramie Shale Oil were mutagenic with the basic fractions being the major contributor of mutagenic activity.

CONCLUSIONS

1. Paraho/Sohio Shale Oil was found to be mutagenic in the Ames assay when assayed with the frameshift strain TA98 and incorporating metabolic activation with rat liver homogenates (Aroclor induced S-9).

2. The mutagenic activity was contributed by the organic constituents of the basic and neutral fractions.

Table IV. Salmonella Mutagenesis of Paraho Shale Oil

	Fractions	% Weight	Specific Activity (Rev/mg)	Weighted Activity (Rev/mg)
1.	Volatiles	0.26	0	0
2.	Et_2O sol. Acids	0.41	0	0
3.	Et_2O sol. Bases	3.32	8000	265
4.	Acid ppt.	0.34	0	0
5.	Base ppt.	0.50	141	1
6.	Total Neutral	95.36	140	133 (116)[a]
	a. Aliphatic	55.61	20	11
	b. Mono + Di Arom.	27.37	0	0
	c. Polyaromatic	6.95	1600	111
	d. Residue	2.15	0	0
			Total:	122
	Total of Base Neutral Fractions:			398 (382)[a]
	ORIGINAL	-	500	500

[a]Obtained by the summation of subfractions.

Table V. Summary of Paraho-Sohio Shale Oil Work With Salmonella System

ORNL Repository No.	Sample Description	Mutagenic Activity (Rev/mg)
4601	Crude Shale Oil	382
4602	Hydrotreated Shale Oil	0
4604	JP-5 Precursor	0
4605	JP-8 Precursor	0
4606	DFM Precursor	0
4607	Hydrotreated Residue	0
4608	JP-5 Product	0
4609	JP-8 Product	0
4610	DFM Product	0
4614	JP-5 Petroleum	0
4616	DFM Petroleum	0

Table VI. Yeast Mutagenesis/Paraho-Sohio Shale Oil

TEST SAMPLE (mg/ml)		S-9	$CAN^r/10^7$ surv. $\pm$ 2. S.D.	Survival
CONTROL		+/-	24.1 $\pm$ 9.6	100%
4601.	Crude Shale Oil	-	42.3 $\pm$ 12.7	74%
		+	26.0 $\pm$ 10.0	69%
	Et_2O sol. Base	-	98.2 $\pm$ 19.4	4%
		+	83.8 $\pm$ 17.9	14%
4602.	HDT Shale Oil	-	18.9 $\pm$ 8.5	100%
		+	21.8 $\pm$ 9.2	72%
	Et_2O sol. Base	-	14.8 $\pm$ 7.5	45%
		+	20.1 $\pm$ 8.8	89%

Table VII. Comparative Studies with Petroleum Crudes and Coal Derived Oils

Repository No.	Sample	% Distribution Acid	Base	Neutral	Mutagenicity (Rev/mg)
5301	Wilmington Crude Oil	0	0	100	5
5305	Recluse Crude Oil	0	0	100	6
1701	SRC-II Fuel Oil	0	36	64	959
1601	H-Coal Distillate	0	41	59	311
1604	H-Coal, HDT (high)	0	0	0	0
4101	Laramie Shale Oil	0	56	44	146
4601	Paraho-Sohio Shale Oil	0	69	30	382
4602	Paraho-Sohio Shale Oil, HDT	0	0	0	0

3. Hydrotreatment of the shale oil abolished the mutagenic activity.

4. Results obtained in the yeast assay supported these observations.

5. Refined shale oil, jet fuel, and diesel fuel marine samples were not mutagenic.

6. The samples rank for their mutagenic activity as coal oils > shale oil > natural petroleum crudes.

REFERENCES

1. B. N. Ames, J. McCann, and E. Yamasaki, Methods for detecting carcinogens and mutagens with the Salmonella/-mammalian-microsome mutagenicity test, Mut. Res. 31, 347-364 (1975).

2. F. W. Larimer, D. W. Ramey, W. Lijinsky, and J. L. Epler, Mutagenicity of methylated N-nitrosopiperidines in Saccharomyces cerevisiae, Mut. Res. 57, 155-161 (1978).

3. T. K. Rao, J. A. Young, A. A. Hardigree, W. Winton, and J. L. Epler, Analytical and biological analyses of test material from the synthetic fuel technologies. II. Extended genetic and biological studies with mutagenic fractions, Mutat. Res. 54, 185-191 (1978).

4. A. P. Swain, J. E. Cooper, and R. L. Stedman, Large scale fractionation of cigarette smoke condensate for chemical and biologic investigations, Cancer Res. 29, 579-583 (1969).

5. J. H. Bell, S. Ireland, and A. W. Spears, Identification of aromatic ketones in cigarette smoke condensate, Anal. Chem. 41, 310-313 (1969).

6. J. L. Epler, M. R. Guerin, and T. K. Rao, Evaluation of feasibility of mutagenic testing of shale oil products and effluents, Environ. Hlth. Perspect. 30, 179-184 (1979).

7. A. R. Jones, M. R. Guerin, and B. R. Clark, Preparative-scale liquid chromatographic fractionation of crude oils derived from coal and shale, Anal. Chem. 49, 1766-1771 (1977).

8. M. R. Guerin, C.-h. Ho, T. K. Rao, B. R. Clark, and J. L. Epler, Polycylcic aromatic primary amines as determinant chemical mutagens in petroleum substitutes, Environmental Research (1980) in press.

9. T. K. Rao, J. L. Epler, M. R. Guerin, B. R. Clark, and C.-h. Ho, Mutagenicity of nitrogen compounds from synthetic crude oils; collection, separation and biological testing. Proceedings of the Second Symposium on the Application of Short-Term Bioassays in the Fractionation and Analysis of Complex Environmental Mixtures, Williamsburg, Va., March 4-7, 1980.

CHAPTER 13

COMPARATIVE MAMMALIAN GENETIC TOXICOLOGY OF SHALE OIL PRODUCTS ASSAYED *InVitro* and *InVivo**

H. Timourian, A. Carrano, J. Carver**, J. S. Felton, F. T. Hatch, D. S. Stuermer, L. H. Thompson, Lawrence Livermore National Laboratory, Biomedical Sciences Division, Livermore, California 94550

INTRODUCTION

The objective of this project is to determine the relative toxicity and mutagenicity (genetic toxicity) of crude and hydrotreated shale oil products from the Paraho surface retort. This was achieved by applying a battery of bioassays emphasizing mammalian systems and having both *in vitro* (cell culture supplemented with microsomes to allow metabolic activation) and short-term *in vivo* components. The results may be compared with the results from bioassays conducted in other laboratories to obtain a basis to estimate health hazards to humans.

A substantial portion of the Biomedical Sciences program at Lawrence Livermore National Laboratory (LLNL) is devoted to the integrated application of cell biology, analytical cytology, and biochemical techniques to problems of environmental mutagenesis, carcinogenesis, and injury to the reproductive system. Most of the bioassays used in this study were developed and validated in our laboratories.

---

*Work performed under the auspices of the U.S. Department of Energy by the Lawrence Livermore National Laboratory under contract number W-7405-eng-48.

**Present address: Battelle-Columbus Laboratories, Biomedical Sciences Division, Columbus, OH 43201.

METHODS

Preparation of Test Materials

The shale oil materials were obtained from the Chemical Repository in 16-oz., brown, borosilicate bottles. The materials were Crude Shale Oil (Repository #4601.13) and Hydrotreated Shale Oil (#4602.13). The bottles were filled up to 85% capacity and flushed with dry argon gas before sealing. The samples of oils were stored at 4°C in a refrigerator where they were protected from light.

Test samples were taken from stock material and applied to the bioassays within 3 days. Before the test samples were removed, the bottles were allowed to warm for 30 minutes after loosening the caps. When the oils became fluid (70-75°F), the caps were tightened, and the fluids were mixed thoroughly by vigorous shaking. Test materials were then removed, the bottles were flushed with dry nitrogen gas, sealed, and cooled to 4°C. The small samples were kept in the dark at 4°C before diluting in either corn oil or dimethyl sulfoxide (DMSO) on the day they were applied to the bioassays.

Organic bases, neutrals, and acids were separated by liquid-liquid extraction and pH adjustments.

Bioassays

Microbial bioassays were used to determine mutation (reversion) induction in histidine auxotrophs (histidine-requiring strains) of *Salmonella typhimurium* as described by Ames.[1] The bacteria were exposed to the shale oil materials, and the revertants that survived and formed colonies in histidine-free media were counted. The number of revertant colonies represented a direct measure of induced mutation frequency. Since some mutagens require enzymatic activation, bacteria were exposed to the oil shale materials with and without S9, a preparation of rat liver microsomes.

Mouse hepatoma cells (line Hepa-1) can be used to detect rapidly benzo(a)pyrene (BaP) and other polycyclic aromatic hydrocarbons in complex mixtures. The Hepa-1 line, which is very sensitive to BaP, and a resistant variant strain,[2] were exposed to the shale oil materials. A rapid screening method based on growth inhibition in multiple well trays was used to assess differential toxicity. Since the major genetic difference between the two strains is the ability to activate BaP, this test does not require the addition of a microsomal activating system.

Chinese hamster ovary (CHO) cells were also used to study cytotoxicity. Plating efficiency at a series of exposure doses was used to determine the D_{37} dose (the dose of an agent that kills 63% of the cells). From a review of the literature on mutagenesis in mammalian cell culture[3] and extensive data of Dr. June Carver in our Division, we found a high correlation between induced mutation frequency (per mol per liter exposure dose) and the D_{37} dose, which was applicable to several genetic markers and in several mammalian species (Figure 1). Thus, to screen and set priorities for further testing, the simple and rapid measurement of cytotoxicity (D_{37} dose) can often predict mutagenicity; any errors will be conservative.

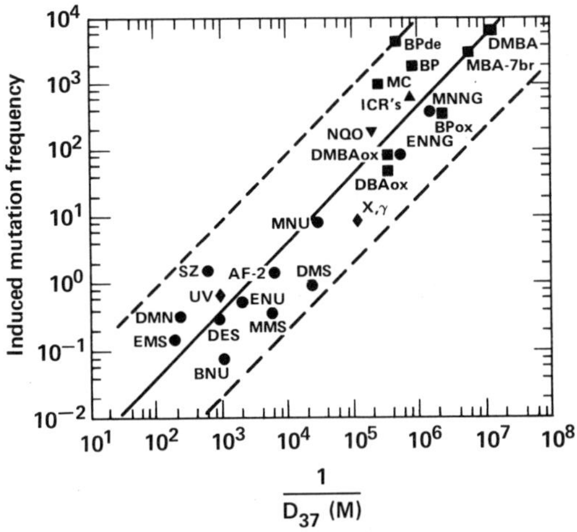

Figure 1. The cytotoxic potency of 22 chemical mutagens correlates with their mutagenic potency as assayed at the *hprt*, *aprt*, and *tk* loci in five rodent and human *in vitro* cell systems. Data on induced mutations are plotted as a function of the reciprocal of D_{37} (M) (the molar dose required to kill 63% of the initial cell population). See Carver *et al*., 1979a[3] for details.

CHO cell mutagenesis. CHO cells were incubated with the test material alone or with a microsomal activating system. After exposure the CHO cells were tested for their reproductive ability by determining plating efficiency and for specific single step mutations at four different loci by determining resistance to four lethal drugs. To determine mutagenicity, cells exposed to the test material were subsequently cultured in lethal concentrations of azaadenine (AA), thioguanine (TG), fluorodeoxyuridine (FUdR) or trifluorothymidine (TFT), or ouabain (OUA). Colony formation in the presence of a drug was taken as a measure of mutation induction by the test material. In general the effects at different markers are correlated, but certain mutagen classes may have differential effects at the markers.[4,5] Thus the multiple marker CHO assay may have a broader sensitivity to detect mutagens than the commonly used cell strains. The CHO cell bioassays are also sensitive to mutation induction by metal ions and their methylation products.[6,7]

Whole mouse toxicity was assessed by determining LD_{50} doses. Juvenile mice (2 weeks old) were injected intraperitoneally with single doses of either Crude or Hydrotreated Shale Oil. We found that the LD_{50} obtained in juvenile mice was similar on a mg per kg basis to the LD_{50} of adult mice.

Mammalian germ cell toxicity was determined by measuring oocyte depletion in 12 day old female mice after *in vivo* exposure to the shale oil materials. Mice were sacrificed 14 days after exposure, their ovaries sectioned and the number of primordial oocytes can be used as a measure of genotoxicity. Oocytes are very sensitive to mutagens, and an accelerated rate of depletion may indicate genetic damage.[8,9] The LD_{50} for γ-rays is about 5 rad, and those for several polycyclic aroatic hydrocarbons range from 1 to 20 mg/kg, which in the newborn mouse represents a total dose of a few micrograms.

Sister chromatid exchanges (SCE) were assayed in mice after *in vivo* exposure to the shale oil materials. The mice were prepared five hours before exposure to the test material, by implanting bromodeoxyuridine (BrdUrd) pellets under the skin. The mice were then sacrificed 20 hours later and cells in their bone marrow examined and scored for SCEs. The SCEs are exchanges of segments between sister chromatids. SCEs indicate that DNA breakage and reunion have taken place. The number of SCEs has been correlated with the number of mutations in mammalian cell cultures for a number of mutagens.[10] We assume that the same relationship will apply in whole animals as in cell cultures.[11]

RESULTS

General Toxicity

From all the bioassay systems that were used in this study, we obtained preliminary information on the general toxicity of the test fractions. The information obtained from the toxicity experiments is used for: (a) a general knowledge of the physiological toxicity of the fraction, (b) preliminary determination of the doses to be tested in subsequent mutagenesis studies, (c) relative toxicity as an indicator of relative mutagenicity, and (d) the determination of whether the toxicity of the test material is masking the mutagenic activity of the fraction.

The Crude and the Hydrotreated Shale Oil fractions are equally toxic to Salmonella strain TA 100 cells (Figure 2). Figure 2 also shows that addition of a microsomal fraction (S9) does not detoxify either the Crude or the Hydrotreated Crude Oils. The toxicity of these fractions to Salmonella TA 100 may be a factor that accounts for the lower mutagenicity in this strain than TA 98.

The toxicity of the shale oil products was also tested on two mouse hepatoma cultured cell lines. The cell lines differed in their ability to metabolize polycyclic aromatic hydrocarbons (PAH). Because of a mutation, the Hepa-1 cell line is less sensitive to benzo(a)pyrene (BaP), it required a 7-day treatment with 3 μg/ml of BaP to become as toxic as 0.1 μg/ml in the parent cell line. The sensitivity to crude shale oil was equal in the two cell lines. However, the cytotoxicity after a 7-day exposure of the cells to the Crude Shale Oil (11 μg/ml) was 8 times higher than exposure to the Hydrotreated Shale Oil (94 μg/ml).

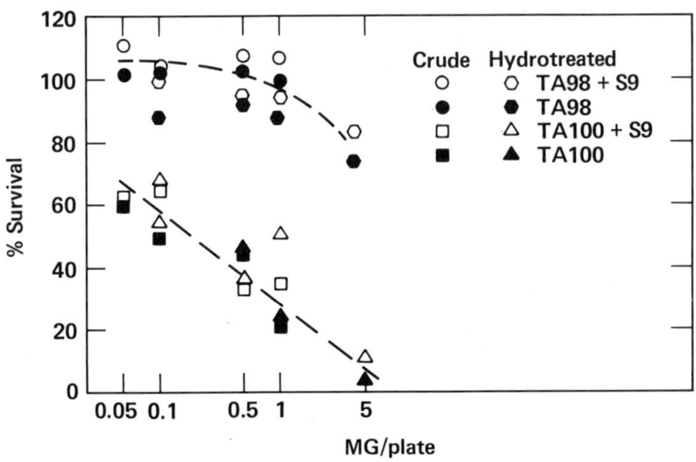

Figure 2. Toxicity to Salmonella strains by Crude and Hydrotreated Shale Oils. Note that S9 prepration has no effect on toxicity and the toxicity to TA 98 is minimal.

The toxicity of the shale oil products was also tested in CHO cells and both the Crude and the Hydrotreated Shale Oils were highly toxic. However, the Hydrotreated Shale Oil was half as toxic as the Crude Shale Oil. The cytotoxicity of both Crude and Hydrotreated Shale Oil was decreased considerably when coincubated with the S9 microsomal fraction. However, the S9 microsomal fraction decreased toxicity presumably by nonspecific absorption since the same effect was brough about when the S9 fraction was devoid of the NADP+ cofactors.

The cytotoxicity of Crude Shale Oil was about three times greater to the whole mouse than the hydrotreated product. The LD_{50} for Crude Shale Oil was 5,000 μg/g body weight; while the LD_{50} for the Hydrotreated Shale Oil was 17,000 μg/g.

Figure 3 compares the toxicity of Crude and Hydrotreated Shale Oils to CHO cells and to the whole mouse. The toxicity to the whole mouse could be compared to the toxicity to CHO cells with added S9 fraction. The absolute amounts of oils that are required for LD_{50} in the whole mouse is about five times greater than required in the CHO cells <u>in vitro</u> if the weight of the mouse is considered equivalent to the volume of a tissue culture solution.

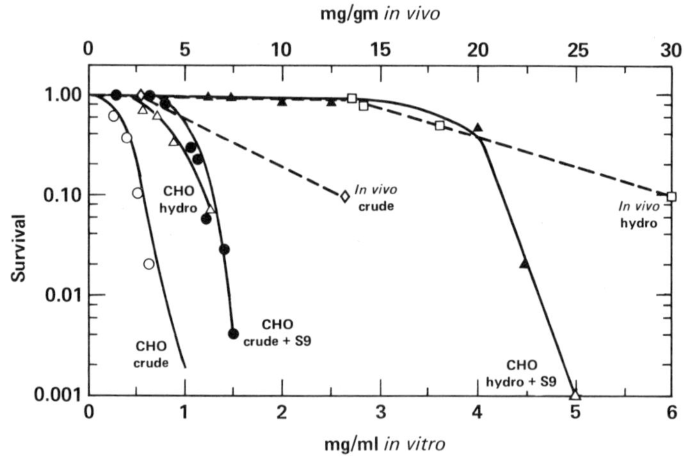

Figure 3. The relative toxicity of Crude and Hydrotreated Shale Oil are compared in CHO cells and the whole mouse.

Genetic Toxicology

The mutagenicity of both Crude and Hydrotreated Shale Oils was tested on two strains of <u>Salmonella</u>: TA 100, which is sensitive to point mutations, and TA 98 which is sensitive to frameshift mutations (Table I). The Crude Shale Oil was mutagenic to both cell strains. Hydrotreatment of the shale oil decreased the mutagenicity in both strains. The drop in mutagenicity differed in the two strains. In TA 100 the mutagenicity dropped to one-fifth and in TA 98, the mutagenicity dropped to one-tenth. When the shale oils were fractionated, the mutagenic activity of the crude oil was found in both the basic and neutral fractions. After hydrotreatment, most of the mutagenic activity disappeared, but some activity remained in the neutral fraction. The acidic fractions never contained mutagenic activity.

Table I. Mutagenicity of Shale Oil to *Salmonella* Strains

Fraction	Revertants per mg/10^8 Salmonella[a] TA 98[b]	TA 100[c]
Crude Shale Oil		
Unfractionated	2470	1220
Basic Fraction	1038	432
Neutral Fraction	1030	650
Sum of Basic and Neutral Fractions	2068	1082
Hydrotreated Shale Oil		
Unfractionated	236	267
Basic Fraction	14	8
Neutral Fraction	176	202
Sum of Basic and Neutral Fractions	190	210

[a]Background revertants have been subtracted.
[b]Sensitive to frameshift mutations.
[c]Sensitive to base-pair substitution mutations.

The mutagenicity of both shale oil products was also tested in CHO cells by determining mutation rates at four loci. The cells were treated with different concentrations of either Crude or Hydrotreated Shale Oil at doses found to allow between one-tenth and one-half survival of the cells. Table II shows that Crude Shale Oil was very weakly mutagenic at two out of the four gene loci tested. All the mutagenicity observed required activation by the S9 fraction. The Crude Shale Oil was mutagenic at the ATPase locus at the highest concentration tested. The frequency of mutants at this locus was tested by determining survival after treatment with the drug ouabain. The Crude Shale Oil was also weakly mutagenic at the thymidine kinase locus. However, the

Table II. Mutagenicity of Crude and Hydrotreated Shale Oil to Four Loci in CHO Cells. Mutagenicity was Dependent on S9-mediated Metabolic Activation and was Lost after Hydrotreatment

Treatment	Dose (μg/ml)	Survival	Mutant frequency/viable cell after drug exposure[a]				
			AA ($x 10^{-5}$)	TG ($x 10^{-5}$)	OUA ($x 10^{-6}$)	TFT ($x 10^{-3}$)	FUdR ($x 10^{-3}$)
Experiment 1							
None	0	1.00	1.6	3.4	7.0	NT	NT
Hydrotreated Oil	4000	0.45	0.4	0.3	1.0	NT	NT
+ S9	4500	0.02	0.2	0.2	0.7	NT	NT
Crude Oil	750	0.84	1.8	1.8	6.7	NT	NT
+ S9	1000	0.28	1.6	2.6	17.0[b]	NT	NT
Experiment 2							
None	0	1.00	2.9	3.4	0.6	2.6	1.4
Crude Oil	1000	0.30	5.6[b]	4.0[b]	1.57	3.8[b]	1.5
+ S9	1100	0.23	3.9[b]	2.1	1.98[c]	5.4[c]	2.1
	1400	0.03	4.2[b]	1.1	1.96[c]	6.1[c]	2.1

[a]AA, azaadenine exposure, mutation at adenine phosphoribosyl transferase locus.
TG, thioguanine exposure, mutation at hypoxanthine phosphoribosyl transferase locus.
OUA, ouabain expoure, mutation at ATPase locus.
TFT, trifluorothymidine, mutation at thymidine kinase locus.
FUdR, fluorodeoxyurindine, mutation at thymidine kinse locus.
[b]Significantly higher than spontaneous rate at 2 SEM.
[c]Higher than spontaneous but not statistically significant. NT = not tested.

mutagenicity was only seen when selection was performed with trifluorothymidine (TFT). Mutation frequencies at the adenine phosphoribosyltransferase and hypoxanthine phosphoribosyltransferase loci were higher at single dose points than spontaneous frequency; however, there was no dose response and the values were not statistically significant. In all experiments the Hydrotreated Shale Oil was not mutagenic.

The induction of sister chromatid exchanges (SCE) was measured in whole mice. As shown in Figure 4, the number of SCEs of mice treated with either Crude or Hydrotreated Shale Oil did not increase significantly at concentrations that were one-third or less of the LD_{50} dose. At higher oil concentrations SCEs could not be determined because cell division was significantly arrested indicating somatic toxicity. Figure 4 also shows that control mutagens like cyclophosphamide and BaP were positive SCE inducers and the induction by BaP was not masked by simultaneous administration of the shale oils.

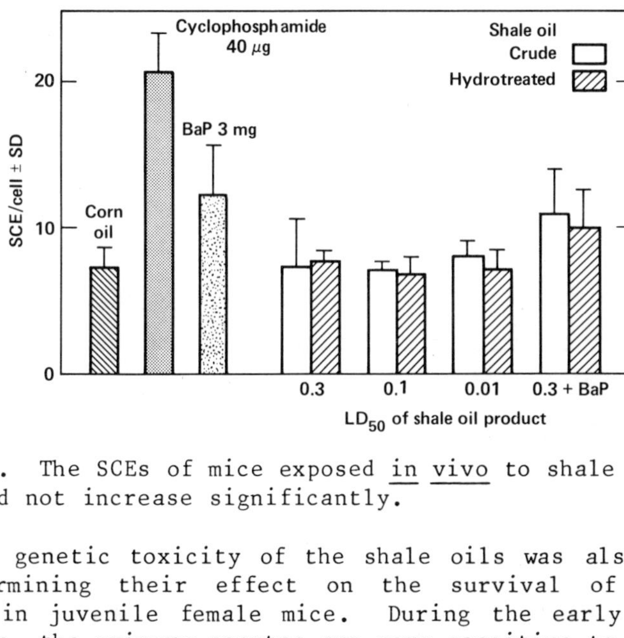

Figure 4. The SCEs of mice exposed *in vivo* to shale oil products did not increase significantly.

The genetic toxicity of the shale oils was also tested by determining their effect on the survival of primary oocytes in juvenile female mice. During the early life of the mouse, the primary oocytes are very sensitive to chemical mutagens.[8,9] Table III shows that the cytotoxicity of Crude Shale Oil to primary oocytes is significant and that it is lost after hydrotreatment of the oil.

Table IV shows a comparison of all the bioassays tested. In this table we give the results of both the cytotoxicity and genetic toxicity studies. The ratio of Crude to Hydrotreated Shale Oil is higher than one in all cases where the crude was active. The ratio varies from one bioassay to the other. The smallest differential sensitivity was in the mouse primary oocyte assay. The largest difference was in the *Salmonella* bioassay, where the Crude Shale Oil was 10 times as mutagenic to the TA 98 strain as the Hydrotreated Shale Oil.

Table III. The Cytoxicity of Crude and Hydrotreated Shale Oil to Primary Oocytes in Young Female Mice

Oil	Oocytes counted	No. Ovaries from Individual Mice Counted	% Survival
Corn (control)	173 ± 27.5	5	100 ± 15.8
Crude Shale	106 ± 2.9[a]	4	59 ± 1.6
Hydrotreated Shale	153 ± 16.9	4	88 ± 9.7

Each tested at LD_{50} for whole mouse.

[a]Significantly different from control at $p < 0.05$.

The results of the present study showing greater toxicity and genotoxicity by the Crude Shale Oil are in agreement with most reports in this symposium and generally support the original results of Coomes,[12] who showed that the carcinogenicity of the shale oil decreased after hydrotreatment. Coomes related the drop in carcinogenicity to a decrease in the nitrogen content of the oil. Our results showing complete loss of mutagenicity in organic bases, but not in the neutral organic fraction, is in agreement with this suggestion.

Our results in this report can also be compared with specific results reported at this meeting. For example, the LD_{50} values for mice (obtained with juvenile mice) are comparable to those obtained by L. H. Smith in adult mice. The LD_{50} for Crude Shale Oil was 5 mg/kg body weight in juveniles and 4 mg/kg in adults, while the LD_{50} for the Hydrotreated Shale Oil was 17 mg/kg in our study and greater than 16 mg/kg in Smith's report. The induction of primary

Table IV. Results of Bioassays of the Cytotoxicity and Genotoxicity of Crude and Hydrotreated Shale Oil

Bioassays	End Point	Crude Shale Oil	Hydrotreated Shale Oil	Crude[a] / Hydrotreated	EMS[b]
Chinese hamster cells	LD_{50} (3-h treatment)				
	-S9	350 μg/ml	800 μg/ml	2.3	675 μg/1ml
	+S9	1,200 μg/ml	3,800 μg/ml	3.7	-
Mouse (whole animal)	LD_{50}	5,000 μg/ml	17,000 μg/ml	3.4	375 μg/g
Hepatoma	Cytotoxicity (7-da)				
	BaP-sensitive	11 μg/ml	94 μg/ml	8.6	(0.1 μg/ml for BaP)
	BaP-resistant	11 μg/ml	94 μg/ml	8.6	(3.0 μg/ml for BaP)
Salmonella	Revertant colonies/μg	10^8 cells			
	TA 98	2.4	0.24	10	-
	TA 100	1.2	0.27	4	1.3
Chinese hamster cells	Mutations at LD_{30} (times background)				
	APRT	1	1	1	75
	HPRT	1	1	1	75
	ATPase	2.5-3	1	2.5	80
	TK	2-2.5	1	2	-
Mouse bone marrow (*in vivo*)	SCE at mouse 0.3 LD_{50}	1	1	1	4
Mouse oocytes	% Survival at mouse LD_{50}	59	88	1.5	5

[a]Because in each bioassay the crude was always more active than the hydrotreated oil, we used used the higher value obtained as the numerator in calculating this ratio.
[b]Ethyl methanesulfonate, a standard mutagen.

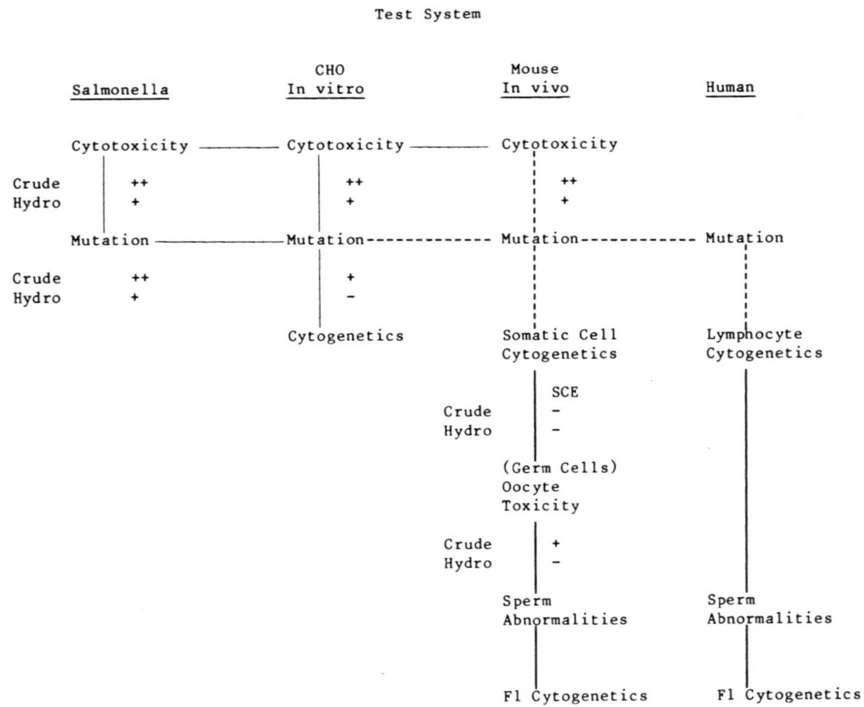

Figure 5. Bioassay data base for risk assessment. Solid lines connect assays which we developed and validated. Broken lines connect assays which are under development. The results from different bioassays on Paraho shale oils, comparing the activity of crude and hydrotreated products are shown as: ++, highly active; +, active; –, non active.

oocyte destruction in the juvenile mouse by Crude Shale Oil was also apparently confirmed by the results reported by M. J. Holland. He discovered ovaries devoid of primary oocytes in adult females after chronic dermal application of shale oils.

Extrapolation of Data from Short-Term Mammalian Genetic Tests to Estimation of Human Risk. The mammalian bioassays should provide an appropriate data base for estimation of risk to man. This risk estimation includes both somatic and germ cell damage as potential indicators of cancer and heritable mutation, respectively. The best approach to quantitative risk evaluation from *in vitro* or *in vivo* data involves the creation of "bridges" both among assays within a test system and also among test systems. The underlying assumption that the response *in vitro* or *in vivo* will apply to the human is necessary unless human samples can be obtained for direct comparison. Figure 5 illustrates how these bridges can be established for one specific battery of bioassays. We also show how the comparative data obtained in this study can serve to start building these bridges.

For example, with this scheme it should be possible to establish the quantitative relation between *Salmonella* revertants and mutation in CHO cells. Within the CHO system the relation between mutations and any of the cytogenic endpoints can be determined. Since the cytogenetic endpoints can be measured *in vitro*, *in vivo*, and in man, this assay can serve as one basis to relate the test data to human exposure. In the absence of human exposure data, it will still be possible to compare quantitatively the several *in vivo* bioassays including those applicable to the F_1 generation. To do this, one would calculate a ratio between the more readily attainable somatic cell data and heritable damage in the F_1. Until such time as human data become available this ratio becomes the best estimate of heritable effects in man. If somatic cell data on the human are available, the risk estimate for the F_1 is considerably improved.

ACKNOWLEDGMENT

We thank P. Berry, S. Healy, E. Salazar, M. Knize, C. Strout, G. Knize, and D. Orwig for technical assistance.

REFERENCES

1. B. N. Ames, J. McCann, and E. Yamasaki, "Methods for Detecting Carcinogens and Mutagens with the *Salmonella*/ Mammalian - Microsome Mutagenicity Test," *Mutat. Res.* 31, 347-364 (1975).

2. O. Hankinson, "Single-step Selection of Clones of a Mouse Hepatoma Cell Line Deficient in Aryl Hydrocarbon Hydroxylase," Proc. Natl. Acad. Sci. U.S. 76, 373-376 (1979).

3. J. H. Carver, F. T. Hatch, and E. W. Branscomb, "Estimating Maximum Limits to Mutagenic Potency from Cytotoxic Potency," Nature 279, 154-156 (1979).

4. June H. Carver, Gerald M. Adair, and Daniel L. Wandres, "Mutagenicity Testing in Mammalian Cells. II. Validation of Multiple Drug-Resistance Markers Having Practical Application for Screening Potential Mutagens," Mutat. Res. (in press); Lawrence Livermore National Laboratory, Livermore, CA, Preprint UCRL-81703 (1979).

5. L. H. Thompson, S. Fong, and K. Brookman, "Validation of Conditions for Efficient Detection of HPRT and APRT Mutations in Suspension-Cultured Chinese Hamster Ovary Cells," Mutat. Res. 74, 21-36 (1980).

6. Robert T. Taylor, June H. Carver, M. Leslie Hanna, and Daniel L. Wandres, "Platinum Induced Mutations to 8-Azaguanine Resistance in Chinese Hamster Ovary Cells," Mutat. Res. 67, 65-80 (1979).

7. Robert T. Taylor, James A. Happe, M. Leslie Hanna, and Rebekah Wu, "Platinum Tetrachloride: Mutagenicity and Methylation with Methylcobalamin," J. Environ. Sci. Health A14(2), 87-109 (1979).

8. R. L. Dobson, C. G. Koehler, J. S. Felton, T. C. Kwan, B. J. Wuebbles, and D. C. L. Jones, "Vulnerability of Female Germ Cells in Developing Mice and Monkeys to Tritium, Gamma Rays, and Polycyclic Aromatic Hydrocarbons," Develop. Toxic. of Energy-Related Pollut. DOE Symp. Series 47, CONF-77107, pp. 1-14 (1978).

9. J. S. Felton, T. C. Kwan, B. J. Wuebbles, and R. L. Dobson, "Genetic Differences in Polycyclic-Aromatic Hydrocarbon-Metabolism and Their Effects on Oocyte Killing in Developing Mice," Develop. Toxic. of Energy-Related Pollut. DOE Symp Series 47, CONF-77107, pp. 15-26 (1978).

10. A. V. Carrano, L. H. Thompson, P. A. Lindl, and J. L. Minkler, "Sister Chromatid Exchange as an Indicator of Mutagenesis," Nature 271, 551-553 (1978).

11. D. G. Stetka, J. L. Minkler, and A. V. Carrano, "Induction of Long-Lived Chromosome Damage, as Manifested by Sister-Chromatid Exchange in Lymphocytes of Animals Exposed to Mitomycin-C," _Mutat_. _Res_. _51_, 383-396, (1978).

12. R. M. Coomes, "Carcinogenic Testing of Oil Shale Materials," 12th Oil Shale Symposium, Colorado School of Mines, Denver, CO, April 18-20, 1979.

CHAPTER 14

TOXICITY OF SHALE OIL TO FRESHWATER ALGAE: COMPARISONS WITH PETROLEUM AND COAL-DERIVED OILS*

J. M. Giddings, Environmental Sciences Division, Oak Ridge National Laboratory, Oak Ridge, Tennessee 37830

INTRODUCTION

Spills of liquid products from oil shale and coal liquefaction may be among the most significant environmental hazards of synthetic fuels production. Full-scale commercialization of coal liquefaction and shale oil production in this country will almost inevitably be accompanied by a certain amount of accidental release of synthetic oils during transportation, storage, and handling.[1] Because oils derived from coal and shale differ from petroleum products in important chemical and physical characteristics, the ecological effects of synfuels spills will be different from, and probably more serious than, the effects of petroleum spills. In recognition of this fact, the Advanced Fossil Energy Program in the Environmental Sciences Division at Oak Ridge National Laboratory is now conducting research on the potential problems of synthetic oil spills. The algal toxicity tests described below are part of this program.

ALGAL TOXICITY TEST

Algae are the dominant primary producers in most freshwater ecosystems. Because algae are at the base of most aquatic food webs, changes in the quantity, quality, or productivity of the algal community could have far-reaching consequences for the rest of the ecosystem. Our algal toxicity screening test is simple, rapid, and ecologically meaningful,

---

*Research supported by the Office of Health and Environmental Research, U. S. Department of Energy, under contract W-7405-eng-26 with the Union Carbide Corporation. Environmental Sciences Division Publication No. 1627.

and it has been used successfully with several dozen aromatic compounds and more than 20 conventional and synthetic oils. The major objective of the test is rapid comparison of different materials with respect to their short-term effects on freshwater algae. The test organisms are _Selenastrum capricornutum_, a unicellular green alga, and _Microcystis aeruginosa_, a non-nitrogen-fixing, blue-green alga. Photosynthetic inhibition is the criterion of toxicity. A second objective of the algal bioassay is determination of the range of toxic concentrations to guide further testing.

The test procedure has been described previously[2,3] and only a brief outline will be presented here. Cells from an actively growing culture are suspended in the test solution and incubated for 4 h. After 2 h, a ^{14}C-bicarbonate solution is added to each sample. The algae take up this inorganic ^{14}C and incorporate it into organic compounds during photosynthesis. At the end of the 4-h incubation, formaldehyde is added to kill the cells. Aliquots from each sample are acidified with HCl to convert all remaining inorganic carbon to CO_2, which is removed by bubbling with air. The remaining (organic) ^{14}C is then assayed by liquid scintillation spectrometry to determine the rate of photosynthesis. Results are expresed as percentages of controls.

MATERIALS TESTED

Coal liquefaction products have been the focus of our most intensive efforts to date.[4,5,6] However, when Dr. Griest informed us of the arrival of the Paraho/SOHIO shale oil suite in the Fossil Fuels Research Materials Facility,[7,8] we requested samples of several of the oils for toxicity screening tests, because of their inherent interest to our program as well as for comparison with petroleum- and coal-derived oils. The oils tested (with their Fossil Fuels Research Materials Facility numbers in parentheses) were the following:

*Crude shale oil (No. 4601)
*Hydrotreated shale oil (No. 4602)
*Hydrotreated residue (No. 4607)
*JP-5 (jet fuel) (No. 4608)
*DFM (diesel fuel marine) (No. 4610)

We also tested three petroleum products for comparison:

*JP-5 (No. 4614)
*DFM (No. 4616)
*No. 6 residual fuel oil (No. 5401).

Crude petroleum was not tested because there is abundant evidence that petroleum crudes are generally less toxic to aquatic organisms than refined petroleum products.[9,10,11]

When oil is spilled on water, most of it either floats on the surface or sinks to the bottom, depending on its density and that of the water. The hazard to aquatic organisms stems more from exposure to components of the oil that dissolve into the water than from direct contact with the oil itself.[11,12] Moreover, the oil may be contained and ultimately recovered, while the water with which it comes in contact will affect a wider area and for a longer period of time. For these reasons and to avoid the experimental difficulties of working with immiscible materials, we tested the water-soluble fractions (WSFs) of the oils rather than the whole oils. Each WSF was prepared by adding oil to distilled water in a 1:8 (oil:water) ratio and stirring very gently for 16 h in the dark. The WSF was then separated from the oil and filtered (Whatman No. 41) before testing. Algal growth nutrients were added, and dilutions were made into fresh algal growth medium.[13] Test solution concentrations were expressed as percentages of full-strength WSF.

RESULTS AND DISCUSSION

The effects of the five oil WSFs on photosynthesis by *Selenastrum capricornutum* are shown in Fig. 1. The Crude Shale Oil WSF was the most toxic; a 10% solution of this material inhibited photosynthesis by nearly 80%. Hydrotreating significantly reduced the toxicity of this oil. The residue of the hydrotreated oil, however, was more toxic than the whole Hydrotreated Oil--in fact, the Hydrotreated Residue was nearly as toxic to *S. capricornutum* as Crude Shale Oil. Neither of the refined shale oil products was toxic in these tests.

The effectiveness of hydrotreating in reducing the toxicity of Crude Shale Oil was demonstrated with both test species (Fig. 2). We believe that this reduction in toxicity is due to removal of nitrogen and oxygen from the oil during hydrotreating.[14] The nitrogen- and oxygen-containing compounds, especially primary amines and phenols, are among the most toxic aromatic compounds to freshwater algae.[3] These compounds are also much more soluble in water than their hydrocarbon analogs, so their abundance in WSFs is greater, proportionally, than their abundance in oil. Preliminary analyses of the WSFs by UV spectroscopy have shown that the concentration of dissolved oil in the Crude Shale Oil WSF is approximately five times that in the Hydrotreated Shale Oil WSF. In studies with coal liquefaction products, we have found the ether-soluble bases to be the most toxic components

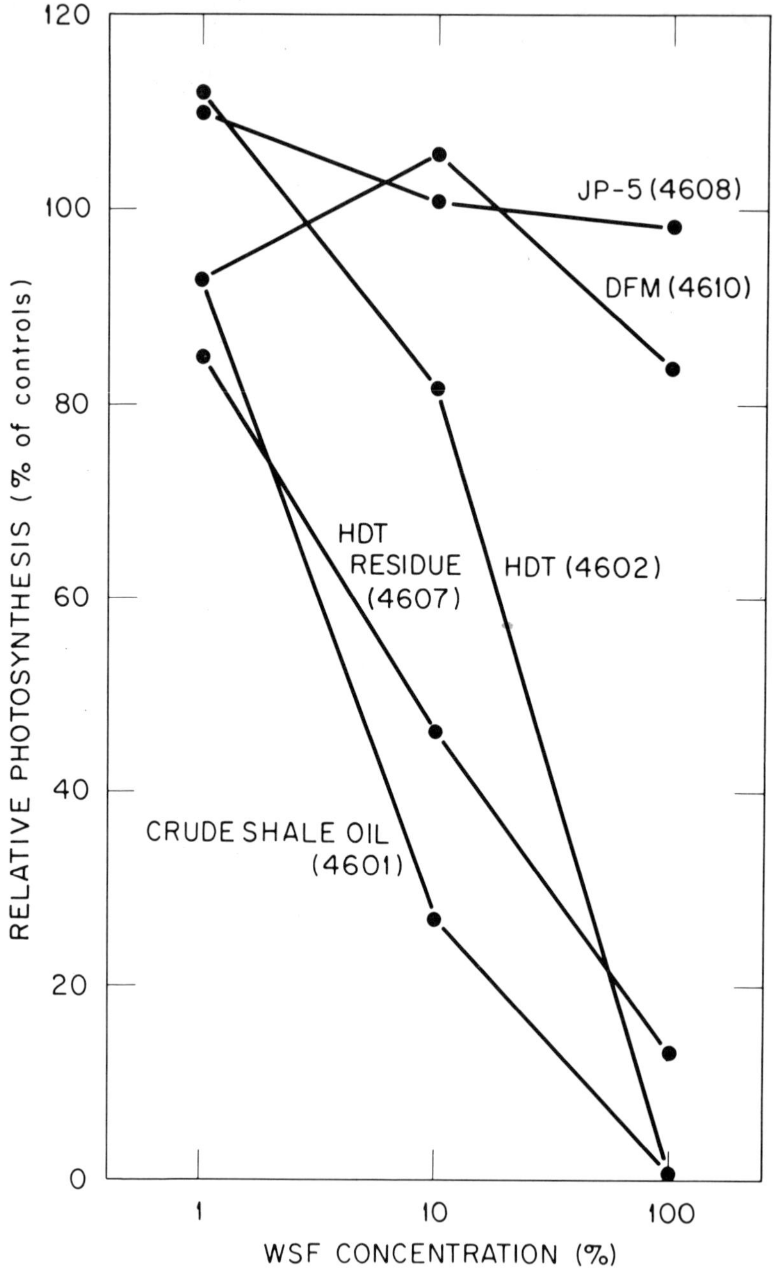

Figure 1. Relative photosynthesis (% of controls) of *Selenastrum capricornutum* exposed to water-soluble fractions (WSFs) of five shale oil materials. Fossil Fuels Research Materials Facility sample identification numbers are given in parentheses.

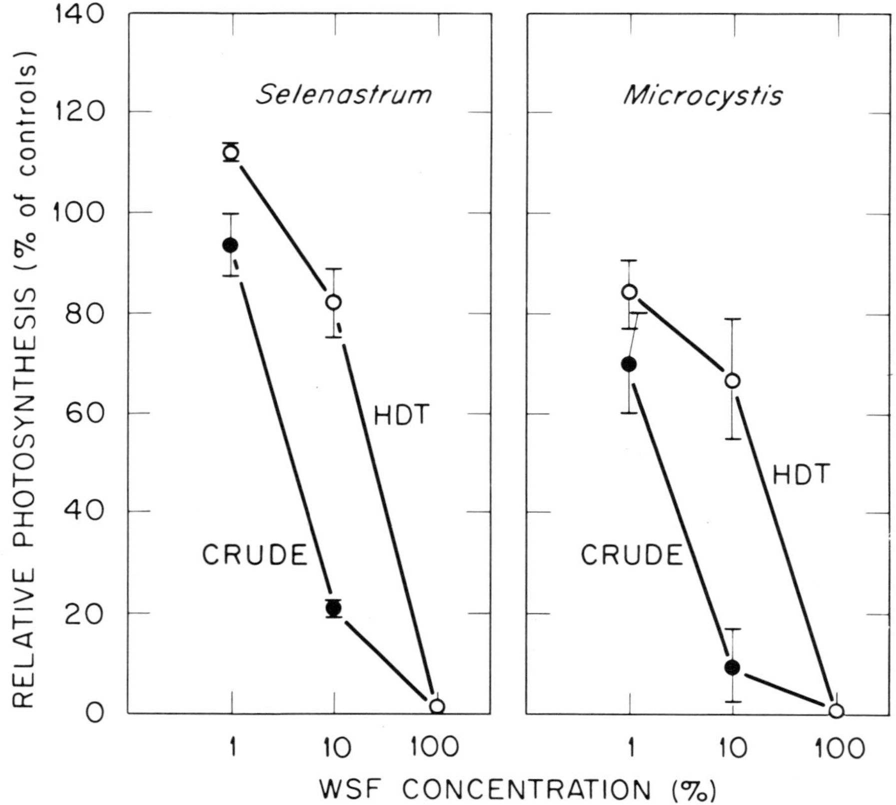

Figure 2. Relative photosynthesis (% of controls) of Selenastrum capricornutum and Microcystis aeruginosa espoxed to water-soluble fractions (WSFs) of Crude and Hydrotreated Shale Oils. Error bars indicate ± 1 S.D.

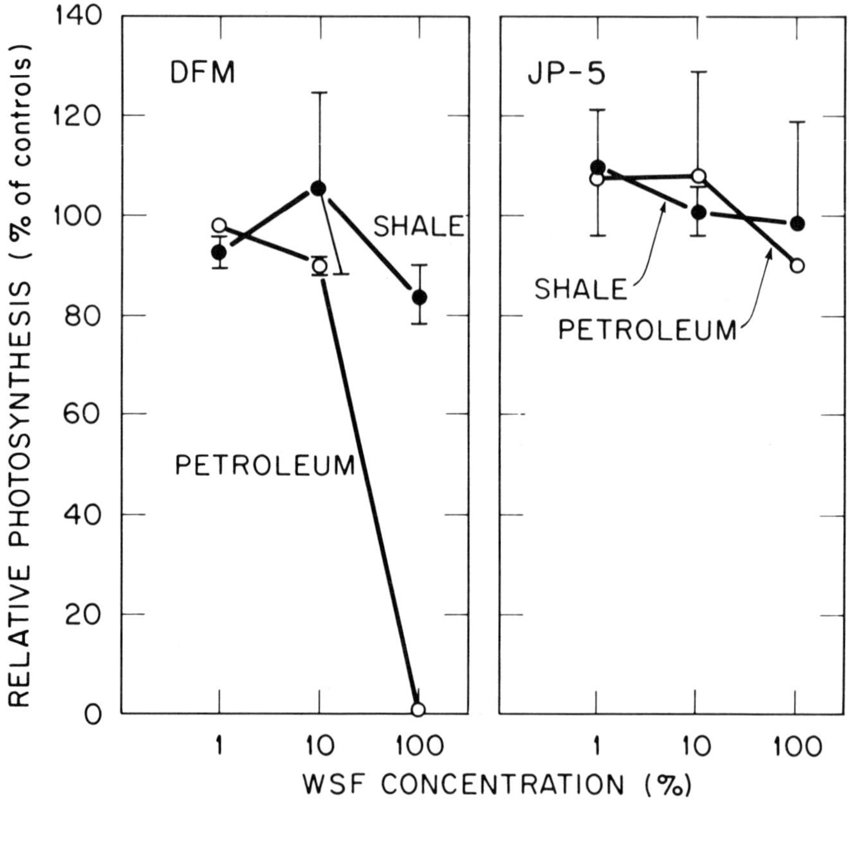

Figure 3. Relative photosynthesis (% of controls) of *Selenastrum capricornutum* exposed to water-soluble fractions (WSFs) of shale oil DFM, petroleum DFM, shale oil JP-5, and petroleum JP-5. Error bars indicate ± 1 S.D.

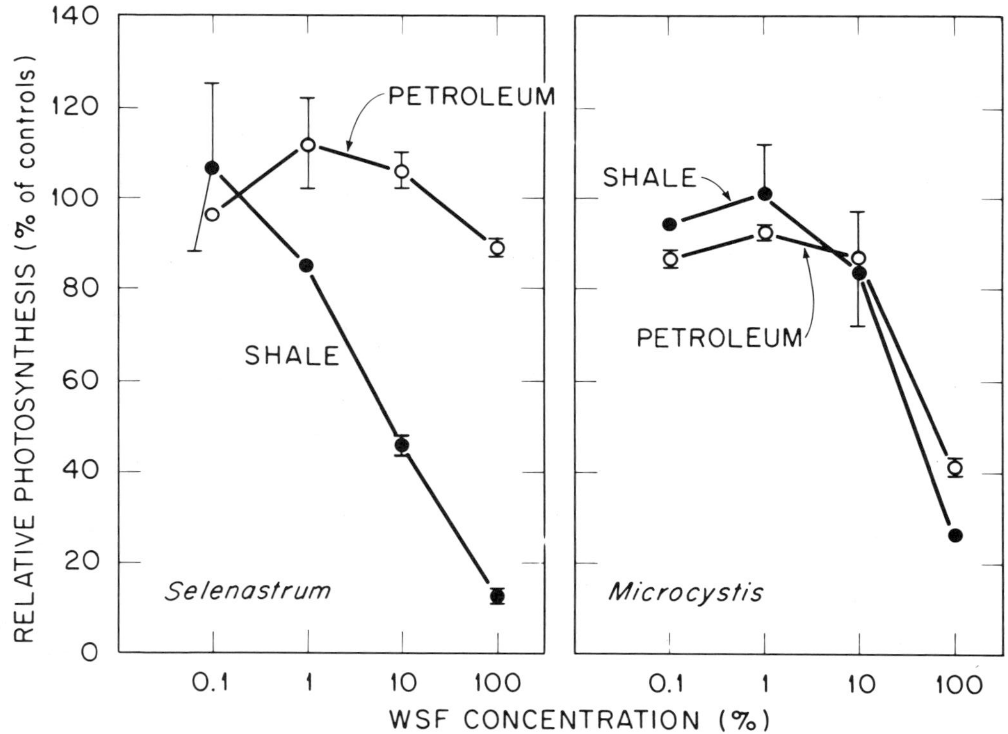

Figure 4. Relative photosynthesis (% of controls) of <u>Selenastrum capricornutum</u> and <u>Mycrocystis aeruginosa</u> exposed to water-soluble fractions (WSFs) of Hydrotreated Residue and a petroleum-derived fuel oil. Error bars indicate ± 1 S.D.

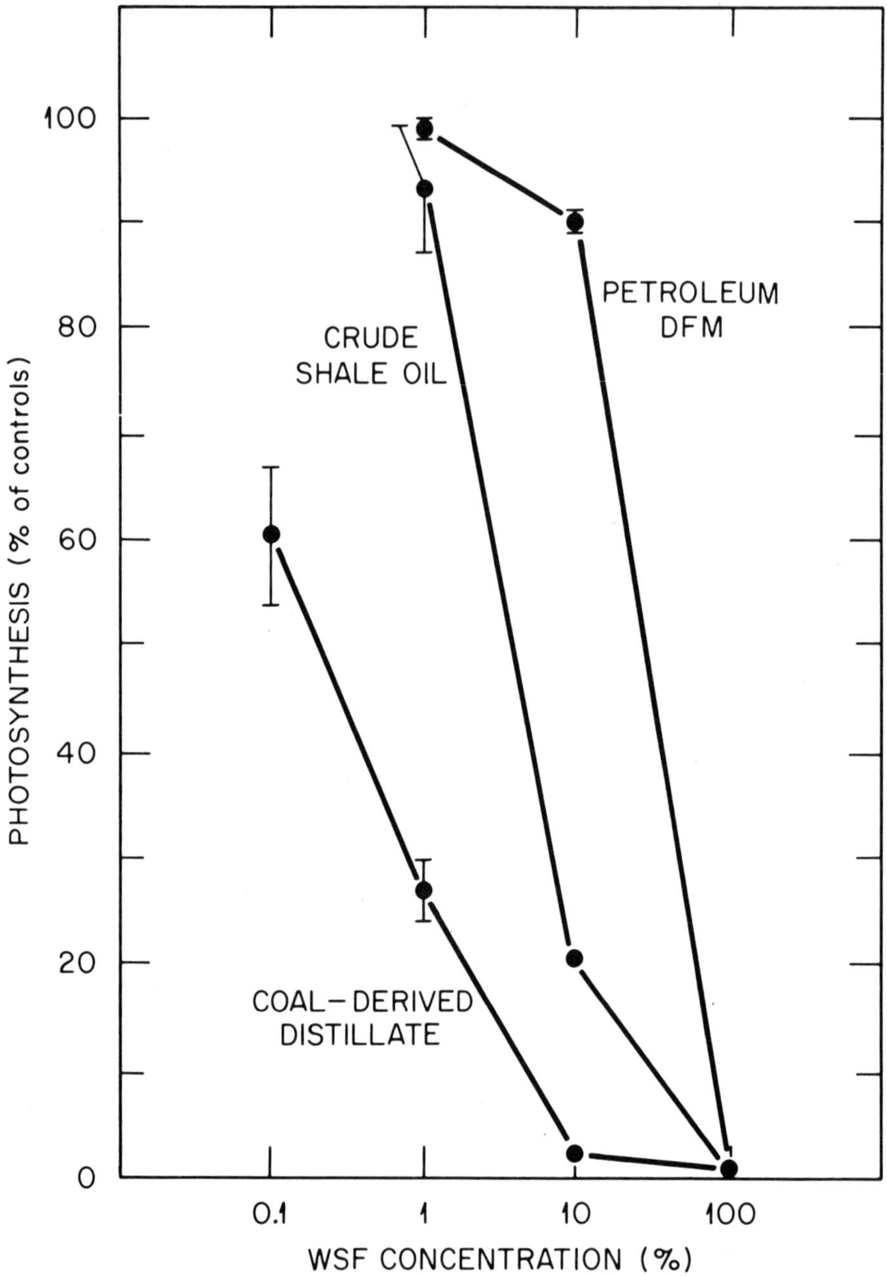

Figure 5. Relative photosynthesis (% of controls) of *Selenastrum capricornutum* exposed to water-soluble fractions (WSFs) of petroleum DFM, crude shale oil, and unrefined coal-derived distillate oil. Error bars indicate ± 1 S.D.

of WSFs, with ether-soluble acids next in importance.[2,15] Similar investigations with shale oils have yet to be carried out, but we expect the same general trends to hold true.

The refined shale oil products, JP-5 and DFM, were not toxic to *S. capricornutum*. The petroleum-derived JP-5 was also nontoxic, but the petroleum DFM was completely inhibitory at 100% WSF (Fig. 3). The DFM was the most toxic petroleum product we have tested in our laboratory.[5,6]

The results with residual fuel oils (Fig. 4) do not present a clear pattern. The petroleum residual fuel oil had little effect on *S. capricornutum*, while the hydrotreated shale oil residue was toxic at 10% WSF. In the case of *M. aeruginosa*, however, the WSFs of the two oils were equally toxic. Subfractionation of these WSFs, followed by bioassays of individual subfractions, would be useful in explaining these results.

Figure 5 presents a comparison of the effects on *S. capricornutum* of the WSFs of petroleum DFM, Crude Shale Oil, and a typical unrefined coal liquefaction product. The 12 coal-derived oils we have tested do not differ greatly in their toxicity to algae; all are considerably more toxic than Crude Shale Oil.[5,6]

CONCLUSIONS

While these results do not constitute a complete evaluation of the relative ecological hazards of the oils tested, several tentative conclusions are suggested:

(1) The WSFs of some of the Paraho/SOHIO shale oils, particularly Crude Shale Oil, are more toxic to algae than WSFs of petroleum products. Shale oil spills might, therefore, be expected to have greater ecological impact than petroleum spills.

(2) Unrefined coal liquefaction product WSFs are more toxic to algae than WSFs of shale oils in the Paraho/SOHIO suite.

(3) Refining reduces the toxicity of shale oil to algae.

REFERENCES

1. N. Leggett, D. Britt, T. Williams, M. Subramanian, and M. Parish, "Spills from the Transportion and Storage of Coal-derived Synthetic Fuel," Oak Ridge National Laboratory, Oak Ridge, TN, ORNL/TM, 1980 (in press).

2. J. M. Giddings, "Four-hour Algal Bioassays for Assessing the Toxicity of Coal-derived Materials," Symposium on Process Mesurements for Environmental Assessment, Atlanta, Georgia, February 25-27, 1980, U. S. Environmental Protection Agency, 1980 (in press).

3. J. M. Giddings, "Acute Toxicity to Selenastrum capricornutum of Aromatic Compounds from Coal Conversion," Bull. Environ. Contam. Toxicol., 23, 360-364 (1979).

4. J. M. Giddings, B. R. Parkhurst, C. W. Gehrs, and R. E. Millemann, "Toxicity of a Coal Liquefaction Product to Aquatic Organisms," Bull. Environ. Contam. Toxicol. 25, 1-6 (1980).

5. J. M. Giddings and J. N. Washington, "Coal Liquefaction Products, Shale Oil, and Petroleum: Acute Toxicity to Freshwater Algae," submitted to Environ. Sci. Technol.

6. L. E. McNeese, "Fossil Energy Program Quarterly Progress Report for the Period Ending September 30, 1979," Oak Ridge National Laboratory, Oak Ridge, TN, ORNL-5612 (1980).

7. W. H. Griest, D. L. Coffin, and M. R. Guerin, "Fossil Fuels Research Matrix Program," Oak Ridge National Laboratory, Oak Ridge, TN, ORNL/TM-7346 (1980).

8. W. H. Griest, M. R. Guerin, L. B. Yeatts, Jr., and B. R. Clark, "Sample Management and Chemical Characterization of the Paraho/SOHIO/U.S. Navy Crude and Refined Shale Oil Suite," this volume.

9. J. W. Anderson, J. M. Neff, B. A. Cox, H. E. Tatem, and G. M. Hightower, "Characteristics of Dispersions and Water-Soluble Extracts of Crude and Refined Oils and Their Toxicity to Estuarine Crustaceans and Fish," Mar. Biol., 27, 75-88 (1974).

10. S. van Gelder-Ottway, "The Comparative Toxicities of Crude Oils, Refined Oil Products and Oil Emulsions," Marine Ecology and Oil Pollution, Halsted Press, 1976, pp. 287-302.

11. S. F. Moore and R. L. Dwyer, "Effects of Oil on Marine Organisms: A Critical Assessment of Published Data," Water Res., 8, 819-827 (1974).

12. D. R. Evans and S. D. Rice, "Effects of Oil on Marine Ecosystems: A Review for Administrators and Policy Makers," Fish. Bull., 72, 625-638 (1974).

13. W. E. Miller, J. C. Greene, and T. Shiroyama, "The Selenastrum capricornutum Printz Algal Assay Bottle Test," EPA-600/9-78-018 (1978).

14. D. Cawein, "Results of the SOHIO Refining Run," this volume.

15. K. E. Cowser, "Coal Liquids Evaluation and Paraho-Sohio Shale Oil Quarterly Progress Report for the Period Ending December 31, 1979," Oak Ridge National Laboratory, Oak Ridge, TN, ORNL/TM-7271 (1980).

CHAPTER 15

COMPARATIVE MUTAGENICITY OF COMBUSTION EMISSIONS OF A HIGH QUALITY NO.2 DIESEL FUEL DERIVED FROM SHALE OIL AND A PETROLEUM DERIVED NO. 2 DIESEL FUEL

J. L. Huisingh[a], D. L. Coffin[a], R. Bradow[a], L. Claxton[a], A. Austin[a], R. Zweidinger[a], R. Walter[b], J. Sturm[b], and R. J. Jungers[a], Environmental Research Center[a], U. S. Environmental Protection Agency, Research Triangle Park, NC 27711, and Transportation Systems Center[b], U. S. Department of Transportation, Cambridge, MA 02142

INTRODUCTION

In the evaluation of the potential toxicity of synthetic fuels, it is a necessity to examine them at each stage during extraction, preparation of the crude fluids, the refining and finally, the end use wherever there is likely to be significant exposure to human beings. While considerable data is becoming available for potential hazards of the earlier stages of production, we know of no available data for combustion products from refined products derived from synthetic fuels. It is, thus, of great utility to look at this problem as soon as such products become available to learn if they differ significantly from analogous petroleum products.

In the case of shale oil, the development of the synthetic crude and its subsequent refining, described elsewhere in this symposium, have presented a timely opportunity to attack, toxicologically, the problem of emission products. The present study involves diesel fuel. Previous studies of combustion emission products of diesel fuel derived from petroleum have demonstrated significant mutagenic activity in the Ames *Salmonella typhimurium* bioassay and in mammalian cells from the organics associated with the exhaust particles.[1,2] Furthermore, concentration of mutagenic activity extractable from exhaust particulate emissions appear to differ according to certain fuel characteristics.[1]

Thus, it is of importance to determine the relative quantity of mutagenic materials contained in diesel fuels from synfuels as compared to those prepared from petroleum.

MATERIALS AND METHODS

Fuels

The fuels used in this study are (1) No. 2 diesel from petroleum, locally obtained from Phillips Petroleum, and (2) diesel fuel marine refined from Paraho crude prepared by Development Engineers, Inc. and refined by Standard Oil of Ohio (SOHIO). See descriptions elsewhere in this symposium. Analytical data for these two fuels are compared to a reference #2 national average (a blend used for vehicle certification) as shown in Table I.

Table I. Fuels Analysis

Analysis	Shale[a]	Petroleum #1[b]	Petroleum #2[c]
API Gravity	38.8	35.5	35.9
Density @ 15°C Kg/liter	0.8359	0.8469	0.8449
% Sulfur	0.024	0.16	0.17
Cetane Index	57.0	48.0	48.7
% Carbon	84.08	84.59	83.12
% Hydrogen	14.96	14.81	14.96
% Nitrogen	<0.01	<0.01	---
Distillation Range			
IBP	395°F	380°F	369°F
10%	450°F	427°F	428°F
50%	503°F	504°F	496°F
90%	553°F	600°F	576°F
EP	574°F	642°F	624°F
FIA			
% Saturates	61.6	66.2	82.4
% Olefins	4.1	1.3	0.4
% Aromatics	34.3	32.5	17.2

[a]Hydrotreated Diesel Fuel Marine courtesy of Navy Shale Oil refining run.

[b]Local #2 Diesel Fuel from Phillips Petroleum, Couch V.

[c]Fuel was Petroleum Derived Diesel Fuel #2, National Average.

Vehicle

The vehicle used in this study was a prototype turbo-charged diesel Volkswagen (VW) Rabbit (European Golf) vehicle of 2,250 lbs inertia weight on loan to the U. S. Department of Transportation (DOT). The vehicle was equipped with a 1.5 liter prototype injection engine with a rated 70 hp at 4,800 rpm. The emissions were measured by DOT as 0.14/.89/1.07 grams per mile of HC/CO/NO_x, respectively. VW also measured the fuel economy as 45.7 mpg urban and 56.5 mpg highway.[3] The particle emission rates measured by DOT were low, varying from 0.05 to 0.3 g/mile for the 15 mph steady-state condition and the 30 mph steady-state condition, respectively.[4] Using the Federal Test Procedure driving cycle, this vehicle emitted 0.17 grams of particles per mile which would meet the newly proposed EPA standard of 0.2 g/mile.

Collection and Extraction of Emission Particles

The vehicle was repetitively operated on a chassis dynamometer to simulate an actual driving pattern, the highway fuel economy test cycle (HWFET). This cycle is 12.75 minutes long over 10.24 miles at an average speed of 48 miles per hour. The particles were collected by filtration on 20-inch by 20-inch Pallflex filters after dilution. The standard dilution tunnel collection technique[1] allows dilution and cooling of the exhaust to obtain more realistic samples of particle bound organics on the two large filters which filter the total tunnel flow. The filters were Soxhlet extracted for 48 hours with dichloromethane. After filtration (0.2 micron filter) and evaporation of the dichloromethane, the samples were diluted in dimethylsulfolxide for bioassay.

Bioassay

The _Salmonella typhimurium_ plate incorporation assay was performed as described by Ames[5] with minor modifications. The modification consisted of the minimal addition of the concentration of histidine directly to the plate media instead of to the overlay and that the plates were incubated for 72 rather than 48 hours. The comparative samples were bioassayed simultaneously in the same experiment for each tester strain.

The studies were performed with triplicate plates at 5 doses with and without metabolic activation. The linear portion of the dose response curves were used to calculate a linear regression line. The slope of the linear regression analysis is reported as revertants/μg of organics tested.

Results

Generally the HC and CO emissions were found to be lower and NO_x levels higher for the shale derived fuel as compared to the petroleum derived fuel.[4] The particle emission rate and fuel economy values were similar with both fuels. The emissions from the shale derived fuel were somewhat higher in extractable organics.

The mutagenic activity of both samples is shown in Table II. The mutagens associated with a wide variety of diesel vehicle emissions show similar characteristics in that they are generally direct-acting mutagens which show responses in tester strains which detect frameshift mutations. The mutagenic activity of the shale derived sample is generally similar to that observed for petroleum derived fuels. TA98 is the most sensitive tester strain for these mutagens and is used here for comparative purposes. In TA98, both samples were more active when tested in the absence of Arochlor induced S9. Although the shale sample demonstrated somewhat higher mutagenic activity, there was no significant difference between the two samples. The shale derived fuel itself should be considered a higher quality fuel than the petroleum derived fuel used in this study based on its higher octane index and lower sulfur content. The nitrogen and aromatic content of the two fuels were not substantially different (Table I).

Table II. Comparison of the Mutagenic Activity of the Organics from Shale vs Petroleum Derived Diesel Fuel Emission in *Salmonella Typhimurium*

Activation System	Slope in Revertants/μg Organics: Petroleum	Slope in Revertants/μg Organics: Shale
	TA100	
-	2.77[a]	2.45[a]
+	1.47	3.51
	TA98	
-	5.10	7.68
+	3.86	4.82

[a]Low R-Square Value
No significant difference between the two samples

Although the mutagenic activity of the organics from the particle emissions was similar for the two fuels, a comparison of the release of mutagens as a result of combustion of these fuels in diesel vehicles requires factoring into the analysis the particle emission rate and percent of extractable organics. The comparative mutagenic emission rate in revertants/mile is shown in Table III. Since the shale sample is somewhat higher both in mutagenic activity and extractable organics, the emission rate in revertants per mile is greater.

Table III. Comparison of the Mutagenic Emission Rate[a] Between a Petroleum and Shale Derived Fuel in *Salmonella Typhimurium* TA98 without Activation

	Rev/μg Org.	% Ext	Rev x 10^5/ gm Part.	Per gm/mi	Revertants/ Mile
Petroleum	5.10	17.8	9.08	0.18	163,000
Shale	7.68	21.4	16.43	0.17	279,000

[a]Turbocharged VW Rabbit diesel vehicle (56.4 mi/gal)

DISCUSSION

The mutagens and potential carcinogens associated with diesel emission are of concern because of the high particle emission rate of diesel vehicles compared to current gasoline vehicles. Mutagenic activity of the particle bound organics appears to be both a function of the type of engine and the fuel quality. Among the various diesel engines tested with the same petroleum derived diesel fuel, a tenfold differnece in the mutagenic activity of the particle bound organics was observed.[6] Within one engine a comparison of four No. 2 diesel fuels showed a sevenfold difference in mutagenic activity with the minimum quality fuel generating higher mutagenicity in the emissions.[1]

Fuel factors are of importance because they could be influenced by the origin of the fuel (petroleum, shale oil coal fluids) or by the type and quality of the refining and upgrading. Of concern here is the fact that the refining and finishing of the diesel fuel marine used in this study appear to have been of high quality as indicated by the chemical data (Table I). Since data from petroleum vary according to the quality, it is pertinent here to point out that the data presented are representative only of this particular shale oil source extraction process and refining method and cannot be applied across the board for shale oil diesel. Until the fuel factors governing mutagenic output by diesel cars are better understood, a great deal of work remains to be done in relating chemical constituents in the fuels to amount of mutagenicity in the emission products. Careful attention to this subject should be rewarding in designing refining methods for both petroleum derived diesels and those from the various synfuel industries including shale oil.

SUMMARY

When shale derived diesel fuel of equal or better quality was compared to petroleum derived fuel, the resulting particle bound organic emissions were not significantly different in mutagenic activity in the Ames *Salmonella typhimurium* plate incorporation assay.

We should expect that a lower quality shale derived fuel may result in the combustion generation of increased amounts of mutagens or mutagenic constituents. Such studies are expected in the future to help delineate the fuels parameters which are critical to the formation of potentially mutagenic combustion products.

REFERENCES

1. J. Huisingh, R. Bradow, R. Jungers, L. Claxton, R. Zweidinger, S. Tejada, J. Bumgarner, F. Duffield, V. F. Simmon, C. Hare, C. Rodriguez, L. Snow, and M. Waters, "Application of Bioassay to the Characterization of Diesel Particle Emissions. Part I. Characterization of Heavy Duty Diesel Particle Emissions. Part II. Application of a Mutagenicity Bioassay to Monitoring Light Duty Diesel Particle Emissions," *Application of Short-Term Bioassays in the Fractionation and Analysis of Complex Environmental Mixtures*, Environmental Science Research, Vol. 15, Ed. by M. D. Waters, S. Nesnow, J. L. Huisingh, S. S. Sandhu, and L. Claxton, Plenum Press, 1979, pp. 382-418.

2. J. L. Huisingh, S. Nesnow, M. Waters, *et al.*, "Application of a Battery of Short-Term Mutagenesis and Carcinogenesis Bioassays to the Evaluation of Soluble Organics from Diesel Particles," In *Proceedings of the International Symposium on Health Effects of Diesel Engine Emissions*, December 1979, in press.

3. A. G. Volkswagenwerk, "Data Base for Light-Weight Automotive Diesel Power Plants," Department of Transportation, Transportation Systems Center, Cambridge, MA 02142, Technical Contract Report No. DOT-TSC-1193.

4. R. A. Walter, S. S. Guayle, and J. C. Sturm, "Fuel Economy and Exhaust Emission Characteristics of a Turbocharged Diesel Automobile," US EPA, Mobile Source Emission Research Branch, Research Triangle Park, NC 27711, Technical Cotnract Report to EPA No. EPA-79-D-X0546.

5. B. N. Ames, J. McCann, and E. Yamasaki, "Methods for Detecting Carcinogens and Mutagens with the _Salmonella_/mammalian Microsome Mutagenicity Tests," _Mut. Res._, _31_, 347-364 (1975).

6. L. Claxton, "Mutagenic and Carcinogenic Potency of Extracts of Diesel and Related Environmental Emissions: _Salmonella typhimurium_ Assay," In _Proceedings of International Symposium on Health Effects of Diesel Engine Emissions_, December 1979, in press.

CHAPTER 16

THE REPAIR DEFICIENT MUTANT *mei-9a* CONFERS HIGH SENSITIVITY ON THE TEST ASSAYING FOR CHEMICALLY INDUCED CHROMOSOME LOSS IN *DROSOPHILA MELANOGASTER**

S. Zimmering, Division of Biology and Medicine, Brown University, Providence, Rhode Island 02912

INTRODUCTION

In view of the projected widespread use of shale oil as a source of energy, shale oil materials demand high priority for mutagenesis testing. For mutagenicity assessment, a battery of test systems should be employed since verification of a positive finding in systems representing a broad spectrum of evolutionary complexity provides the strongest inferential evidence regarding the probable mutagenic potential of the compounds in humans. Strong recommendation can be made for the use of Drosophila as one of the test systems acknowledging its status as a premier *in vivo* test system for the detection of "genic lesions", i.e. gene/point mutations (and small deletions) induced in germ cells by both direct and indirect acting compounds using the classical sex-linked recessive lethal test.[14] However, Drosophila procedures employed in the detection of chemically induced chromosome lesions/breaks, i.e., those leading ultimately to chromosome loss or translocations have proven considerably less than satisfactory. With few exceptions, such procedures have required substantially higher concentration than the recessive lethal test to demonstrate a positive response or have proven wholly negative even at concentrations producing high rates of recessive lethals.[3,13] Accordingly, the development of a sensitive and rapid test in Drosophila to detect chromosome lesions in germ cells would not only enhance substantially the utility of Drosophila as comprehensive screen in assessing the mutagenic potential of suspected mutagens/-carcinogens, reduce substantially the cost of screening compounds in Drosophila for chromosome breakage, but provide a firm basis for the evaluation of Drosophila as a rapid and sensitive higher eukaryote *in vivo* germ cell analogue to mammalian somatic cell cytogenetics.

*Preparation of this manuscript was supported in part by contract No. DE-AC02-80EV1042.A000 from the Department of Energy.

The purpose of the present paper is to review recent findings by Zimmering and co-workers on a radically improved test for cheomosome loss used as an assay to detect induced chromosome lesions.[17-19] This improved procedure has not only yielded significantly higher frequencies of loss with chemical mutagens inducing lower but detectable levels of loss with conventional tests, but, more critically, has permitted detection of highly significant increases in loss with compounds failing to elicit a positive response in traditional tests even at substantially higher concentrations.

The modified test for chromosome loss consists primarily of substituting for an ordinary repair proficient Pl female, a strain of female homozygous for the excision repair $mei\text{-}9^a$ mutant.[1] The rationale for the initial trials in our laboratory with $mei\text{-}9^a$ females derived from the following observations: (a) the hypersensitivity of $mei\text{-}9^a$ larvae to X-ray and UV-induced killing resulting from a defect in repair replication;[4,8] (b) the significantly higher rates of spontaneous chromosome breakage/interchange in $mei\text{-}9^a$ somatic cells;[2] (c) the well documented observation that repair systems in the female may influence the steps involved in the finalization of breaks (or pre-break lesions) induced in treated sperm;[7,9,16] and (d) direct evidence from this laboratory of a marked potentiation in the ability to detect chromosome loss induced in the paternal genome by X-rays when Pl $mei\text{-}9^a$ females are employed compared with ordinary repair proficient females.[5] Finally, it is of importance to note that the test for chromosome loss requires only an Fl generation, i.e., definitive scoring is made on Fl progeny. In contrast, the conventional test for translocations requires an F2 and often an F3 generation for completion. Accordingly, our efforts have been directed toward seeking improvements in the former procedure.

MATERIALS AND METHODS

Four compounds have thus far been analyzed in this regard. They include methyl methanesulfonate (MMS), dimethylnitrosamine (DMN), diethylnitrosamine (DEN) and procarbazine (natulan). Males, 0-8 h old, carrying the ring (closed) X chromosome, R(1)2, marked with *y* (yellow body) and a doubly marked Y chromosome, B^SYy^+, were aged 24 h, starved 6 h, and then permitted to feed for a given length of time on the test solution in 10% sucrose (treated series) or on a solution of 10% sucrose alone (non-treated series). Information on concentrations and length of treatments is given below.

For treatments lasting more than 24 h, solutions were renewed after each 24 h period. Following treatment, males

Compound	Conc. in mM	Length of Treatment (hours)
MMS	0.5	48
	1.0	48
DMN	2.5	24
DEN	10	24
	5	72
Procarbazine	10	48

were mated with females for 3 days (♀ Brood A). Males were discarded and females transferred to fresh food for an additional 3 days (♀ Brood B) to determine if enhancement of effects in ♀ Brood A could be achieved in this second 3-day storage period; females were then discarded. (In the case of many compounds, storage of sperm in the female for longer periods of time results in an increase in break-related events likely due to the production of potential breaks requiring time to develop into actual breaks.[10])

Illustrative of the definitive P1 matings is the example shown in Figure 1. Current crosses are made of

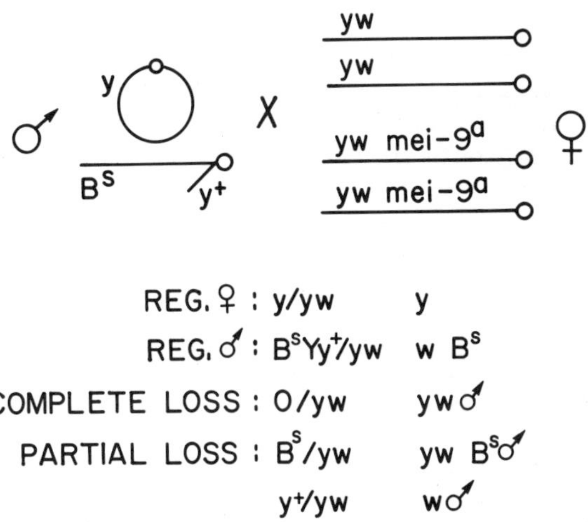

Figure 1. Phenotypes of F1 progeny expected from matings of R(1)2,y/B^SYy$^+$ males with y w repair proficient females and y w mei-9^a females including regular females and males, and products of complete loss of X or Y and partial loss of the Y chromosome.

R(1)2,y/B^SYy$^+$ males (treated or not) with y w ordinary, repair proficient females, and with y w mei-9^a females. In the legend are given the F1 phenotypes representing complete loss of the ring-X or the Y chromosome, and partial loss of the Y chromosome reflecting loss of either the y$^+$ or the B^S marker. An additional point should be noted. Under the conditions of the experiment, the ring-X is ordinarily transmitted to the F1 regular female and the Y chromosome to the F1 regular male. Accordingly, preferential non-recoverable loss of the ring-X chromosome resulting in the death of the ring-X-bearing females is expected to cause a shift in the sex ratio (Regular ♀♀/Regular ♂♂) in favor of males. To obtain an estimate of the extent of the shift in sex ratio and thereby a measure of inferred ring loss, the formula 1-[SR(T)/SR(NT)] X 100 was employed where SR = sex ratio, T = treated series, and NT = nontreated series. Values arrived at thereby for each experiment are shown in the last column of the tables as "Ring Loss (SR)".

RESULTS AND DISCUSSION

Data are presented in Tables I-IV. Attention is directed to the histograms (Figs. 2-5) which summarize the data regarding frequencies of induced complete loss, partial loss and complete + partial loss for each of the four compounds tested. Major points of interest in comparing results with ordinary and mei-9^a females are noted below in connection with each of the compounds.

MMS (Fig. 2, Table I): (a) all induced frequencies in the mei-9^a series represent significant increases above the non-treated mei-9^a control ($P < .001$); (b) induced frequencies of all categories of loss were substantially higher with mei-9^a than ordinary females in both ♀ Brood A and B and at both concentrations ($P < .001$); and (c) at a concentration of 0.5 mM, results for all categories of loss with ordinary females were negative in contrast with the positive response observed at this concentration with mei-9^a females.

DMN (Fig. 3, Table II): (a) all induced frequencies with mei-9^a females reflect significant increases above mei-9^a controls ($P < .001$); and (b) induced frequencies of all categories of loss were considerably higher with mei-9^a vs. ordinary females in both ♀ broods ($P < .001$); and (c) no evidence of induced partial loss was found with ordinary females, but highly significant increases ($P < .001$) with mei-9^a females. It should be noted that preliminary data from this laboratory clearly suggest that at a concentration of 1.25 mM (one-half that reported herein), response with mei-9^a females is positive but wholly negative with ordinary females.

Table I. Fl progeny from matings of R(1)2*y*/B^SYy^+ males, treated with MMS or Not, with Repair Proficient (+) females or *mei-9*^{a} females. Males were treated with 0.5 mM or 1.0 mM MMS for 48 hours.

Conc. mM	Time (h)	♀	T, NT	♀ Bd.	Reg. ♀♀	Reg. ♂♂	CL	PL	N	%CL	%PL	%CL+PL	SR	% Ind. CL	% Ind. PL	% Ind. CL+PL	% Ring Loss (SR)
0.5	48	*mei-9*	NT	A	909	777	15	0	1701	0.88	0.00	0.88	1.17				
				B	1035	971	16	1	2023	0.79	0.05	0.84	1.07				
			T	A	3437	4475	183	27	8122	2.25	0.33	2.59	0.77	1.37	0.33	1.71	34.2
				B	1490	2719	84	23	4316	1.95	0.53	2.48	0.55	1.16	0.48	1.64	48.6
		y w	NT	A	1898	1650	26	0	3574	0.73	0.00	0.73					
				B	1507	1389	32	2	2930	1.09	0.07	1.16					
			T	A	1988	2006	31	0	4025	0.77	0.00	0.77	0.99	0.04	0.00	0.04	13.9
				B	1912	1900	43	0	3855	1.12	0.00	1.12	1.01	0.03	-0.07	-0.04	6.5
1.0	48	*mei-9*	NT	A	3507	3243	72	2	6824	1.05	0.03	1.08	1.08				
				B	3050	3232	66	4	6352	1.04	0.06	1.10	0.94				
			T	A	2177	3096	137	36	5446	2.52	0.66	3.18	0.70	1.47	0.63	2.10	35.2
				B	334	706	41	13	1094	3.75	1.19	4.94	0.47	2.71	1.13	3.84	50.0
		y w	NT	A	4059	3550	49	0	7658	0.64	0.00	0.64	1.14				
				B	3555	3098	38	0	6691	0.57	0.00	0.57	1.15				
			T	A	4470	4301	80	0	8851	0.90	0.00	0.90	1.04	0.26	0.00	0.26	8.8
				B	3827	3691	92	3	7613	1.21	0.04	1.25	1.04	0.64	0.04	0.68	9.6

Notes: T = treated series; NT = nontreated series; Reg ♀♀ = regular females; Reg ♂♂ = regular males; CL = complete loss of a sex chromosome in the male; PL = partial loss of the Y chromosome (loss of y^+ or B^S); SR = sex ratio; Reg. ♀♀/Reg. ♂♂. %Ring Loss (SR) = deficiency in Fl regular females ring-bearing females relative to Fl regular Y bearing males used as a measure of loss of the ring-X and computed as given in the text.

Table II. Fl progeny from matings of R(1)2,*y*/B^SYy^+ males, treated or not with 2.5 mM DMN, with repair proficient (+) females or *mei-9*^{a} females.

Conc. mM	Time (h)	♀	T, NT	Bd.	Reg. ♀♀	Reg. ♂♂	CL	PL	N	%CL	%PL	%CL+PL	SR	% Ind. CL	% Ind. PL	% Ind. CL+PL	% Ring Loss (SR
2.5	24	*mei-9*	NT	A	7987	7115	159	6	16257	1.04	0.04	1.08	1.12				
				B	8673	8290	190	2	17155	1.11	0.01	1.12	1.05				
			T	A	2697	3678	130	30	6535	1.99	0.46	2.45	0.73	0.95	0.42	1.37	34.8
				B	468	857	43	24	1392	3.09	1.72	4.81	0.55	1.98	1.71	3.69	47.6
		y w	NT	A	5012	4674	83	1	9770	0.85	0.01	0.86	1.07				
				B	5636	4836	68	2	10541	0.65	0.01	0.65	1.17				
			T	A	7808	7165	114	2	15089	0.76	0.01	0.77	1.09	-0.09	0.00	-0.09	-1.9
				B	6496	6106	163	3	12768	1.28	0.02	1.30	1.06	0.63	0.01	0.65	9.4

Notes: See Table I.

Table III. Fl progeny from matings of R(1)2,y/$B^{S}Yy^{+}$ males, treated or not with DEN, with repair proficient (+) females or mei-9a females. Males were treated with 10 mM DEN for 24 h or 5 mM DEN for 72 h.

Conc. mM	Time (h)	♀	T, NT	♀ Bd.	Reg. ♀♀	Reg. ♂♂	CL	PL	N	%CL	%PL	%CL+PL	SR	% Ind. CL	% Ind. PL	% Ind. CL+PL	% Ring Loss (SR)
10	24	mei-9	NT	A	3214	2861	66	4	6145	1.07	0.07	1.14	1.12				
				B	3562	3669	86	3	7320	1.17	0.04	1.22	0.97				
			T	A	4517	6346	291	49	11203	2.60	0.44	3.03	0.71	1.53	0.37	1.89	36.6
				B	3862	5899	217	42	10020	2.17	0.42	2.58	0.65	1.00	0.38	1.36	33.0
		y w	NT	A	1740	1489	32	2	3263	0.98	0.06	1.04	1.16				
				B	1846	1691	36	0	3573	1.01	0.00	1.01	1.09				
			T	A	2375	2101	39	2	4517	0.86	0.04	0.91	1.13	-0.12	-0.02	-0.13	2.6
				B	3075	2930	53	0	6058	0.87	0.00	0.87	1.05	-0.14	0.00	-0.14	3.7
5	72	mei-9	NT	A	1551	1512	48	0	3111	1.54	0.00	1.54	1.03				
				B	1759	2118	49	2	3928	1.25	0.05	1.30	0.83				
			T	A	2510	3972	217	23	6722	3.23	0.34	3.57	0.63	1.69	0.34	2.03	38.8
				B	2549	4741	187	47	7524	2.49	0.62	3.11	0.54	1.24	0.57	1.81	34.9
		y w	NT	A	2941	2558	58	1	5567	1.04	0.02	1.06	1.15				
				B	4843	4435	85	1	9364	0.91	0.01	0.92	1.09				
			T	A	4381	3852	87	4	8324	1.05	0.05	1.10	1.14	0.01	0.03	0.04	0.87
				B	6639	6489	130	1	13259	0.98	0.01	0.99	1.02	0.07	0.00	0.07	10.5

Notes: See Table I.

Table IV. Fl progeny from matings of R(1)2*y*/*B*^{S}Y*y*$^{+}$ males, treated with 10 mM procarbazine for 48 h, with repair proficient (+) females or *mei-9*^{a} females.

Conc. mM	Time (h)	♀	T, NT	♀ Bd.	Reg. ♀♀	Reg. ♂♂	CL	PL	N	%CL	%PL	%CL+PL	SR	% Ind. CL	% Ind. PL	% Ind. CL+PL	% Ring Loss (SR)
10	48	*mei-9*	NT	A	1714	1669	41	0	3424	1.20	0.00	1.20	1.03				
				B	1879	2132	45	0	4506	1.11	0.00	1.11	0.88				
			T	A	672	1200	72	16	1960	3.67	0.82	4.49	0.56	2.47	0.82	3.29	45.6
				B	574	1107	40	18	1739	2.30	1.04	3.34	0.52	1.19	1.04	2.23	41.1
		y w	NT	A	1381	1182	26	0	2589	1.00	0.00	1.00	1.17				
				B	1338	1171	25	0	2536	0.99	0.00	0.99	1.14				
			T	A	1524	1516	27	0	3067	0.88	0.00	0.88	1.01	-0.12	0.00	-0.12	13.7
				B	1833	1481	27	0	3341	0.81	0.00	0.81	1.24	-0.18	0.00	-0.18	-8.3

Notes: See Table I.

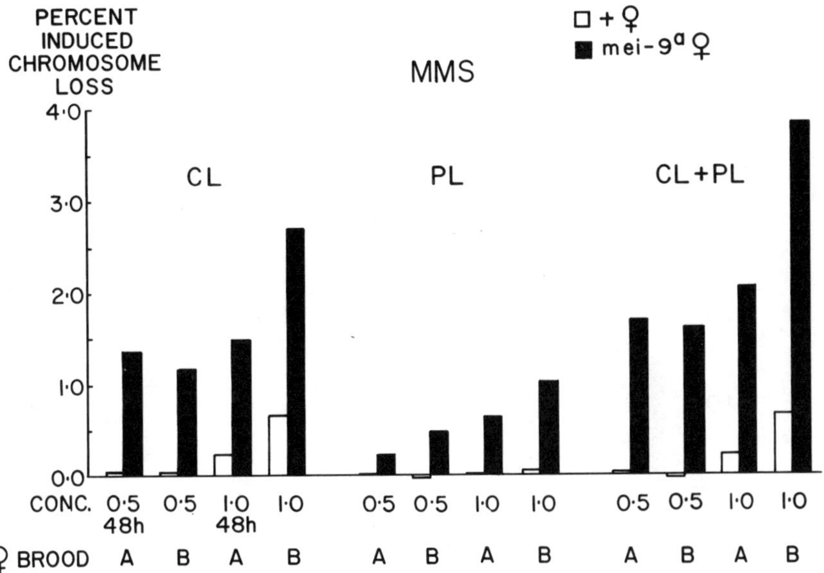

Figure 2. Induced complete loss (CL), partial loss (PL) and complete + partial loss (CL+PL) in F1 progeny of R(1)2,y/B^{S}Yy^{+} males treated with 0.5 mM or 1.0mM MMS for 48h and repair proficient (+) females or mei-9^{a} females.

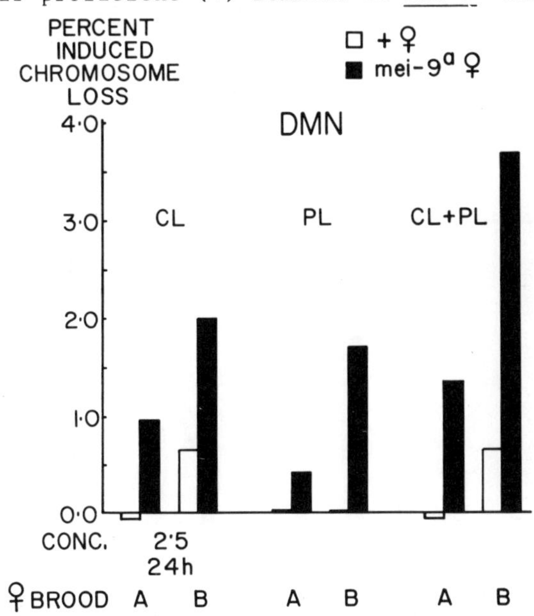

Figure 3. Induced complete loss (CL), partial loss (PL) and complete + partial loss (CL+PL) in F1 progeny of R(1)2,y/B^{S}Yy^{+} males treated with 2.5 mM for 24h and repair proficient (+) females or mei-9^{a} females.

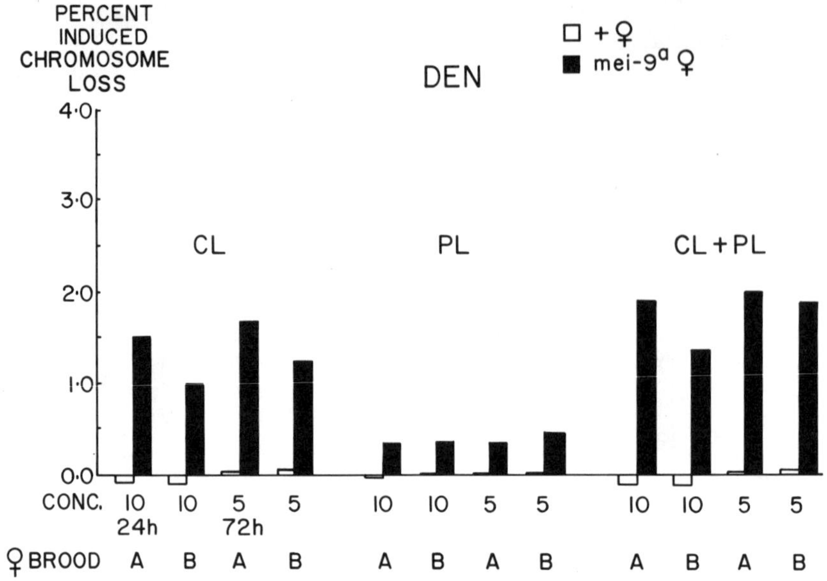

Figure 4. Induced complete loss (CL), partial loss (PL) and complete + partial loss (CL+PL) in F1 progeny of R(1)2,_y/B_^{S}Y_y_$^+$ males treated with 10mM DEN for 24h and 5mM DEN for 72h and repair proficient (+) females or _mei-9_^{a} females.

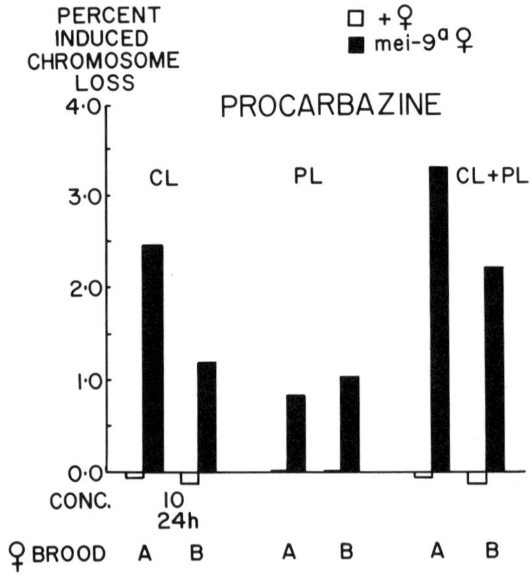

Figure 5. Induced complete loss (CL), partial loss (PL) and complete + partial loss (CL+PL) in F1 progeny of R(1)2,_y/B_^{S}Y_y_$^+$ males treated with 10mM procarbazine for 48h and repair proficient (+) females or _mei-9_^{a} females.

The remaining two compounds, DEN and procarbazine, were selected deliberately for testing since a large body of data from experiments with ordinary females by Vogel and co-workers provided no evidence that either compound induced chromosome loss (with ordinary females) or translocations even at doses producing high frequencies (10-20%) of sex-linked recessive lethals.[3,13] As is shown below, both compounds prove positive for chromosome loss with mei-9^{a} females.

DEN (Fig. 4, Table III): With treatments of 10 mM for 24 hours or 5 mM for 72 hours, a significant increase above controls for each category of loss was observed with mei-9^{a} females ($P < .001$). Evident in the figure, however, is the wholly negative response with repair proficient females.

Procarbazine (Fig. 5, Table IV): As with DEN, substantial increases above non-treated controls were observed with mei-9^{a} females ($P < .001$) but no evidence of a positive response with ordinary females. Accordingly, failure to find a positive response on chromosome loss with DEN or procarbazine with ordinary females confirms earlier findings by Vogel and co-workers in this regard.

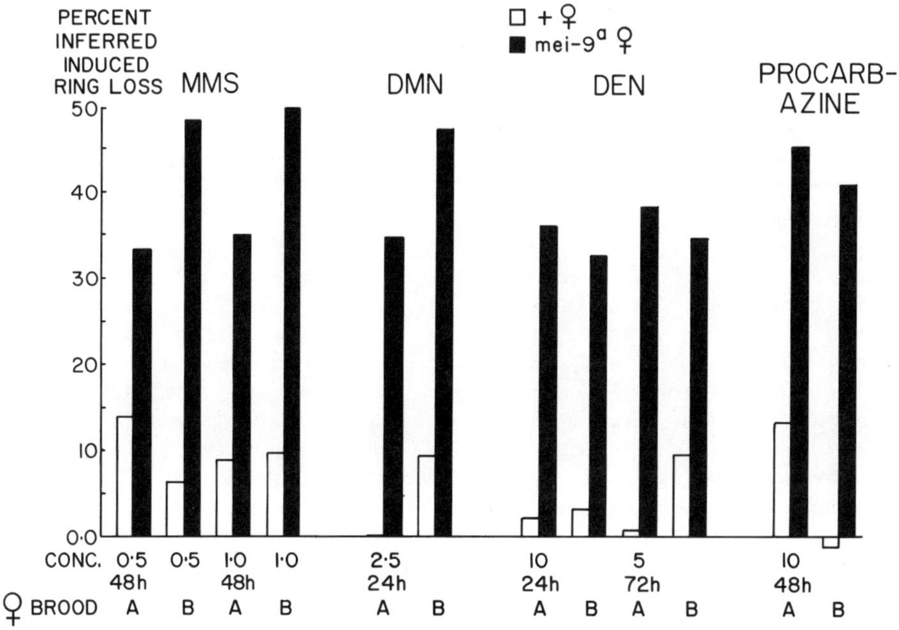

Figure 6. Inferred induced loss of the ring-X [R(1)2] based on shifts in sex ratio following treatment with MMS, DMN, and procarbazine.

A summary of inferred induced ring-X loss for each compound is shown in Figure 6 (as given in the last column of each of Tables I-IV). The strong potentiating effect of $\underline{mei\text{-}9^a}$ females is self-evident.

Although the utility of this protocol measured against the conventional translocation test cannot be assessed directly as yet since the genetic makeup of the strains employed in experiments with $\underline{mei\text{-}9^a}$ were not suitable for parallel testing of F1 males for translocations, a tentative comparison is presented in Table V between data compiled thusfar in this laboratory with $\underline{mei\text{-}9^a}$ on recovered chromosome loss (complete + partial loss) and translocation test data provided in the literature from experiments carried out in E. Vogel's laboratory. Estimates of spontaneous translocation frequencies in the literature range from 0.01% to 0.05%.[7,12] Concentrations used for translocation tests most closely approximating those employed with $\underline{mei\text{-}9^a}$ were selected. The last two columns under Chromosome Loss headed "N-Required" represent the sample size required to achieve a 90% or 99% certainty of detecting a "true" difference between the frequency in the treated series (T) and the non-treated series (NT) at the 5% level of significance.[11] Simple inspection of the table makes it clear that with the exception of MMS where detection capabilities of the two tests are similar (bearing in mind, however, that the translocation test requires 1-2 additional generations for completion), the $\underline{mei\text{-}9^a}$ test for chromosome loss is far more sensitive, most dramatically illustrated in the case of procarbazine. Of interest in connection with the latter is that studies with procarbazine in mouse bone marrow cells provide strong evidence that the compound induces chromosome breakage in mammalian cells.[6]

Accordingly, given evidence of a substantial induction of chromosome lesions leading to chromosome loss following treatment of Drosophila sperm/spermatids with MMS, DMN, DEN and procarbazine and matings with $\underline{mei\text{-}9^a}$ females, it is clear that damage induced in chromosomes of the paternal genome by these compounds must be far greater than deducible from crosses with ordinary repair proficient females. This inference suggests, in turn, that the evidence for the poor chromosome breaking ability of compounds such as DEN and procarbazine, despite high frequencies of recessive lethals, is explainable not primarily as due to some unique insensitivity of Drosophila chromosomes to breakage,[14] but to the remarkable ability of the ordinary Drosophila female to correct chromosome lesions (premutational or otherwise). Finally, it is of obvious importance to note that the modified protocol described herein requires validation using a large number of compounds representing different chemical

Table V. Comparison of frequencies of induced heritable translocations using conventional strains (Vogel and co-workers) and induced chromosome loss using *mei-9*[a] females (Zimmering and co-workers).

	Translocations						Chromosome Loss mei-9[a]							
											% CL+PL		N Required	
Compound	Conc. (mM)	Duration (h)	Sperm Stored (days)	#T	N	%T	Conc. (mM)	Duration (h)	Sperm Stored (days)	Table** ♀ Bd.	T	NT	0.90	0.99
MMS	2.0	24	6	13	334	3.89	1.0	48	6	IB	4.94	1.08	280	489
DMN	2.0	27	15	2	896	0.22*	2.5	24	6	IIB	4.81	1.12	391	684
DEN	5.0	72	21	0	1713	0.00	5.0	72	6	IIIA	3.57	1.54	973	1702
	10.0	72	7	1	629	0.16*								
Procarb-azine	20.0	48	7	0	3814	0.00	10.0	48	6	IVA	4.49	1.20	486	849
	30.0	48	10	0	2139	0.00								

*Not significant.
**This paper.
[a]Ref. 15
[b]Ref. 13.
[c]Ref. 3.
[d]Ref. 18, and Tables I and IV, this paper.
[e]Ref. 17, and Table II, this paper.
[f]Ref. 19, and Table III, this paper.

Duration = length of treatment
#T = number of translocations
N = number of translocation tests
%T - % translocation

classes. It would be desirable as well to establish the lowest concentration of a compound producing a positive response in this system and in parallel test for sex-linked recessive lethals. In this way, a far more reliable comparison between recessive lethal and chromosome break induction in Drosophila may be achieved.

REFERENCES

1. B. Baker and A. Carpenter, "Genetic Analysis of Sex Chromosomal Meiotic Mutants in *Drosophila melanogaster*," *Genetics* *71*, 255-286 (1972).

2. B. Baker, J. Boyd, A. Carpenter, *et al.*, "Genetic Controls of Meiotic Recombination and Somatic DNA Metabolism in *Drosophila melanogaster*," *Proc. Natl. Acad. Sci.* (USA) *73*, 4140-4144, (1976).

3. W. Blijleven and E. Vogel, "The Mutational Spectrum of Procarbazine in *Drosophila melanogaster*," *Mutation Res.* *45*, 47-59 (1977).

4. J. Boyd, M. Golino and R. Setlow, "The *Mei-9^{a}* Mutant of *Drosophila melanogaster* Increases Mutagen Sensitivity and Decreases Excision Repair," *Genetics* *84*, 527-544 (1976).

5. S. F. Cooper and S. Zimmering, "A Genetic Study of the Effect of the *Mei-9^{a}* Mutation in Drosophila on Spontaneous and X-ray Induced Paternal Sex Chromosome Loss," *Mutation Res.* (in press).

6. J. Heddle and W. Bruce, "Comparison of the Micronucleus and Sperm Abnormally Assays for Mutagenicity and Carcinogenic Activities of 61 Different Agents," Cold Spring Harbor Symposium, *Origins of Cancer*, September 7-14, 1976.

7. D. Mendelson and F. H. Sobels, "The Inhibitory Effect of Caffeine on Maternal Repair of Radiation-Induced Breaks in Drosophila," *Mutation Res.* *26*, 123-128 (1974).

8. T. Nguyen and J. Boyd, "The Meiotic *Mei-9* Mutants of *Drosophila melanogaster* are Deficient in Repair Replication of DNA," *Mol. Gen. Genet.* *158*, 141-147 (1977).

9. C. Osgood and S. Zimmering, "Effects of Caffeine on Maternal Repair Systems in *Drosophila melanogaster*. Concentration-Dependent Reversals on the Effects of Caffeine on Chromosome Loss and Autosome-Autosome Translocations Induced by X-rays in the Paternal Genome," *Mutation Res.* *63*, 79-86 (1979).

10. H. Slizynska, "Cytological Analysis of Storage Effects on Various Types of Complete and Mosaic Change Induced in Drosophila Chromosomes by Chemical Mutagens," Mutation Res. 19, 199-213 (1973).

11. R. Sokal and F. Rholf, Biometry, W. H. Freeman and Co., San Francisco, 1969.

12. Traut, H., "The Linear Dose Dependence of Radiation-Induced Translocation Frequency in Drosophila melanogaster at relatively low X-radiation doses. Int. J. Radiat. Biol 7, 401-403 (1963).

13. E. Vogel and B. Leigh, "Concentration-Effect Studies with MMS, TEB. 2,4,6-TriCl-PDMT and DEN on the Induction of Dominant and Recessive Lethals, Chromosome Loss and Translocations in Drosophila perm," Mutation Res. 29, 382-395 (1975).

14. E. Vogel and F. H. Sobek, "The Function of Drosophila in Genetic Toxicology Testing," in Chemical Mutagenesis: Principles and Methods for Their Detection, A. Holleander, Ed., p. 93, Plenum Press, NY, 1976.

15. E. Vogel and A. Natarajan, "The Relation Between Reaction Kinetics and Mutagenic Action on Mono-Functional Alkylating Agents in Higher Eukaryotic Systems. I. Recessive Lethals and Translocations in Drosophila," Mutation Res. 62, 51-100 (1979).

16. F. Wurgler and P. Mair, "Genetic Control of Mutation Induction in Drosophila melanogaster. I. Sex Chromosome Loss in X-rayed Mature Sperm," Mutation Res. 15, 41-53 (1972).

17. S. Zimmering, A. W. Hartmann and S. Cooper, "Evidence that the Rapair Deficient Mei-9 a Female in Drosophila melanogaster is a Strong Potentiator of Chromosome Loss Induced in the Paternal Genome by Dimethylnitrosamine," Environ. Mutag. (in press).

18. S. Zimmering and K. L. Kammermeyer, "Potentiation of Chromosome Loss Induced in the Paternal Genome by Methyl Methanesulfonate and Procarbazine Following Matings with Repair Deficient Mei-9[a] Female in Drosophila melanogaster," Environ. Mutag. (in press).

19. S. Zimmering, A. W. Hartmann and A. W. Song, "The Repair Deficient Mei-9[a] Drosophila Female Potentiates Chromosome Loss Induced in the Paternal Genome by Diethylnitrosamine," Mutation Res. (in press).

CHAPTER 17

TREATMENT OF MULTIVARIATE ENVIRONMENTAL AND HEALTH PROBLEMS ASSOCIATED WITH OIL SHALE TECHNOLOGY*

M. J. Kland, Energy and Environment Division, Lawrence Berkeley Laboratory, Berkeley, California 94720

INTRODUCTION

As the cost of foreign oil derived from traditional sources skyrockets, the economics of developing rich native oil shale deposits of the Green River Basin have become more attractive. Unfortunately, the potential occupational and environmental risks of large scale shale oil production are sufficient to cause concern. In addition, assessment of these risks presents a complex logistic problem. Thus, a preliminary listing of only those identified chemical byproducts found in oil shale and shale oil by EMIC and ETIC, has already yielded some 500 organic compounds, 40 metals and other inorganic species.[1] Included among these substances are potential and known carcinogens. Existing biological data are often inconclusive or inadequate and the logistics and costs of assessing potential environmental and health (E&H) effects by traditional biological methods are prohibitive.

Currently *in vitro* test methods such as the Ames Salmonella tests for mutagenicity are used by the EPA in conjunction with such *in vivo* biological test data as are available to establish a ranking of compounds for future *in vivo* testing of suspect commercial chemicals. A similar approach is being used on shale oil fractions which consist of complex mixtures of neutral, basic and acidic classes of organics. However, *in vitro* testing cannot yet replace long term animal test procedures--themselves often not conclusive because of variations in metabolic pathways with species. The Ames test is not reliable for metals--an important contaminant of shale oils and retort waters.[2a,b]

*Work done under the auspices of the Department of Energy under Contract No. W-7405-eng-48.

Under the best circumstances, there remain unanswered questions of synergisms--the interactions of substances to enhance or inhibit bio-effects; questions of the reliability of mutagenicity as a ranking criterion for testing such a wide variety of substances as are found in raw and spent shales, shale oils and retort waters; and questions also of the relationships, if any, between carcinogens, mutagens, and teratogens.

Since the complex environmental and health problems posed by commercial oil shale development are clearly multi-factorial, their solution will require a statistical matrix approach whose components are described in the following sections.

THE MATRIX APPROACH

In the complex chemical environment typical of most oil shale samples, the matrix approach involves parameters which relate biological activity directly or indirectly to chemical composition and ultimately to molecular structure. In addition to distinctly structural criteria such as rings, double bonds and functional groups, thermodynamic data may also provide useful predictive parameters. Such general properties as spectral and photodynamic behavior, molecular conformation and electro- and nucleophilicities have all been associated with carcinogenicity.[3,4,5]

On the biological side, quantitative mutagenic responses, mammalian cellular transformations, DNA repair and DNA sedimentation analysis are among the useful parameters for correlation.

An overview of matrix methodology applied to the examination of repository samples for potential health effects is given by Coffin (this volume, Chapter 1).

For complex chemical mixtures, quantitative structure-activity and molecular connectivity relationships (QSAR, MC) within classes of compounds, where available, are useful functions, particularly when applied in conjunction with the methods of factor analysis (FA) and pattern recognition (PR). In the remainder of this chapter, each of these four methodologies will be considered as it relates to the assessment of environmental and health effects in multivariate, multicomponent systems similar to those encountered in the Oil Shale Repository studies reported at this Symposium.

STRUCTURE-ACTIVITY RELATIONSHIPS (SAR)

Historical

The relationship between chemical structure and reactivity of organic compounds was recognized long ago by Crum Brown and Fraser[6a,b] who used the following equation to express biological response (R) in terms of chemical structure (C):

$$R = f(C). \tag{1}$$

However it was Hammett[7] who gave structure-activity relationships of chemical systems more precise definition:

$$\log K_\sigma = \log K_\mu + \sigma\rho. \tag{2}$$

Here K's are rate or equilibrium constants, K_σ refers to a substituted benzoic acid, and K_μ is the unsubstituted parent acid. Rho (ρ) and σ are reaction and substitutent constants, respectively.

Still later Hansch[8] and coworkers used the substituent constant π, related to the octanol-water distribution coefficient P, to further quantify these relationships:

$$\pi_\chi = \log P_\chi - \log P_H. \tag{3}$$

Here P_H and P_χ are the partition coefficients of the parent compound (H) and its substituted derivative (χ). This function and other versions of it have been widely used in drug and pesticide studies.

In 1973 Fahmy[9] and coworkers used the Taft steric parameter E_s[10] to relate LD_{50}s of DDT derivatives to the size of a substituent on one of the benzene rings:

$$\log LD_{50} = \alpha + \beta E_s^x + \gamma [E_s^x]^2. \tag{4}$$

Figure 1 shows a typical curve obtained in these studies, indicating that there is an optimum substituent size for maximum LD_{50}, hence the quadratic form of the equation and the resemblance of this plot to a typical potential energy curve.

A number of approaches to SAR have been used. For a recent review of the parameters and methodologies of quantitative structure-activity relationships (QSAR) the reader may consult the chapter by Osman, et al in Ref. 11.

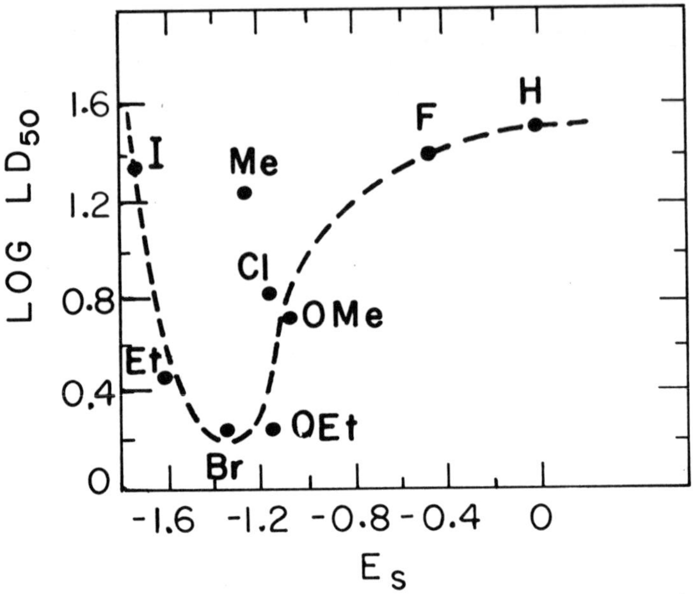

Figure 1. Relationship between toxicity and E_s for 1,1,1-trichloro-p-methyl-p'-x-diphenylethanes.

Polynuclear Aromatic Hydrocarbons (PAH)

Polynuclear aromatic hydrocarbons (PAH) are found in the neutral fraction of shale oils. They are among the substances present in shale oils and retort waters which bear close scrutiny, since there are a number of known potent carcinogens among them. It is therefore of interest to consider two simple features, namely ring number and position, as they relate to a health effect: carcinogenicity. In Figure 2 a series of PAHs related to anthracene (I) and phenanthrene (II) are shown. The parent compounds are listed as "inactive". However, it is important to remember that the supposedly inactive lower members of a PAH series may function as initiators, co-carcinogens or synergists to enhance carcinogenic activity of the whole over the sum of its parts. Also at very high doses an occasional papilloma or tumor has been noted at the site of application of the inactive compound itself.

Fusion of a fourth aromatic ring in the 1,2 or [a] position of anthracene enhances carcinogenicity only slightly (III). However, the asymmetric dibenz[a,h]anthracene (IV) is a moderately active carcinogen. The corresponding linear pentacene (not shown) is also classified as inactive.

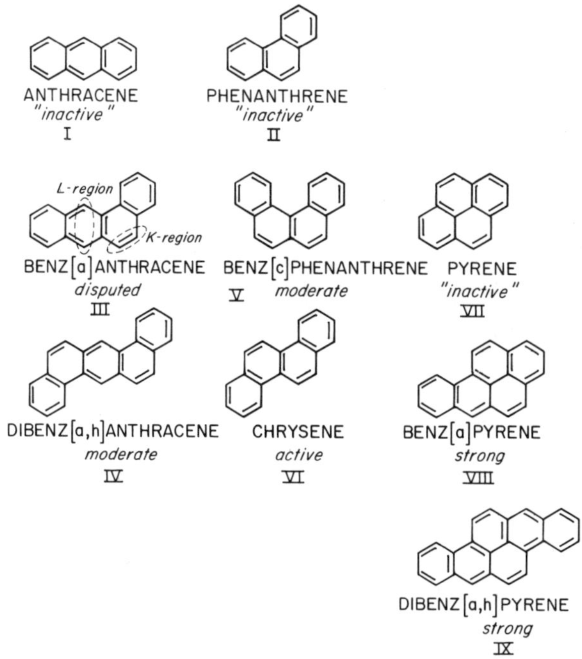

Figure 2. Carcinogenic activities of polynuclear aromatic hydrocarbons.

Early work on the carcinogenic activities of PAHs pointed to the importance of high double bond activity at the K-region as an essential feature (see structure III). Availability of the L-region which is active toward 1,4-addition appeared to diminish carcinogenic activity, at least in the lower fused ring systems. Substitution of methyl groups in the L-region positions have an activating effect, presumably because of the positive inductive (electron releasing) effect of the $-CH_3$ group. Thus, 7,12-dimethylbenzanthracene is a potent carcinogen.

In view of the varied possible mechanisms of carcinogenesis, and the many factors affecting the biological activity of a chemical compound, it is not surprising that no absolute generalizations relating structure to carcinogenicity have emerged. However, many of the more potent PAH carcinogens are relatively good electron donors, have low ionization potentials, form charge-transfer complexes with ease and exhibit "photodynamic activity"--e.g., behave as photochemical sensitizers.

Other structure-activity relationships which have been used with some success include molecular size and thickness and the relationship of PAH structures to those of the steroids. Perhaps the most universal attribute of the carcinogenic PAHs is their structural relationship to phenanthrene (II). This is consistent with theoretical arguments requiring a region of high double bond activity (K-region).[12]

More recent computer-assisted studies of a PAH data set using pattern recognition techniques by Yuan and Jurs[13] confirm the importance of multidimensional analysis in applying SAR to the prediction of carcinogenicity of PAHs. Among the important determinants mentioned were molecular geometry, structural characteristics, lipophilicity and steric effects.

Aromatic Compounds Containing Nitrogen

Aromatic amines and their alkyl derivatives are prominent contaminants found in the base fractions of shale oils and retort waters. Aniline, all three toluidines and five of the xylidines have been reported, as have some of the mixed ethyl methyl derivatives.[1]

Historically aniline itself was viewed as a suspect carcinogen, because of the clustering of bladder cancers observed by Rehn (1895) in the Swiss dye industry, where commercial aniline was the starting material for magenta and other dyes. Later it was shown that the actual causative agents were aniline derivatives and the 1- and 2-naphthylamines (Figure 3). By the early 1900's bladder cancer was a recognized occupational disease wherever an established chemical industry flourished.[14]

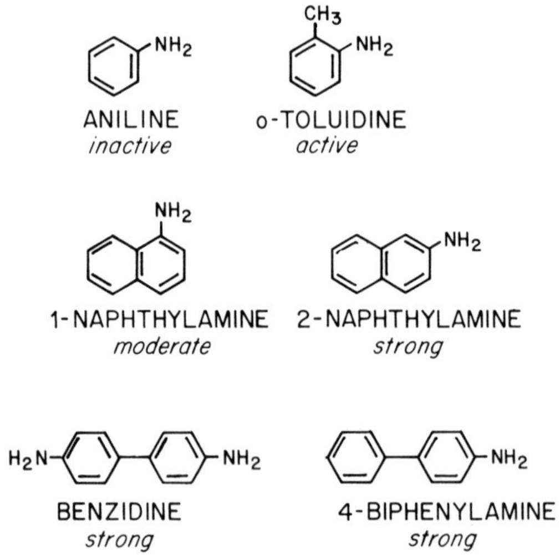

Figure 3. Carcinogenic activities of aromatic amines.

Aromatic amines show much greater species specificity than the PAHs and, unlike the latter, do not induce tumors locally at sites of application. Instead the bladder, liver, or intestines may be affected. Also, 2-naphthylamine, a powerful bladder carcinogen in humans and dogs, is essentially inactive in both rabbits and rats. This highly specific behavior implies that arylamines require metabolic activation to render them carcinogenic. Current evidence is that Nhydroxylation of the substituted arylamine must occur, possibly followed by esterification of the OH group, yielding an unstable intermediate. The latter then breaks down to a positive nitronium ion, an electrophile capable of reacting at a nucleophilic (electron rich) site of the cell (Figure 4).

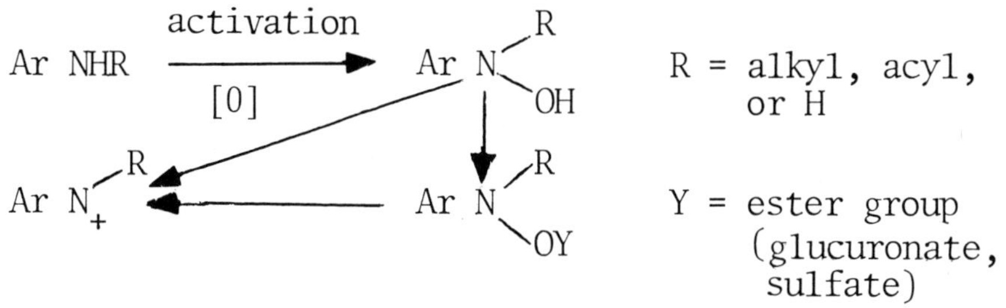

Figure 4. Possible metabolic activation pathways for aromatic amines.

Although aniline itself does not appear to be a carcinogen in man, o-toluidine (hydrochloride) is, and other derivatives of aniline have also induced cancers in mammals. In extended anilines such as 4-biphenylamine, the position of the amino group is important. In general aromatic amines with a conjugated para substituent are much more active than their isomers with a free para position. Similarly the substitution of a methyl group ortho to an arylamino group often increases carcinogenicity.

With rare exceptions (e.g., o-toluidine) carcinogenicity data of the many isomeric arylalkylamines found in shale oils are missing, incomplete or equivocal. This is unfortunate because recent mutagenicity testing of shale oil fractions indicates that the organic base fraction, because of its magnitude, may prove to be a greater health hazard than the PAHs of the neutral fraction which are present in lesser amounts than encountered in other synfuels and coal.

Among the other known tumor initiators found in shale oils are the benzacridines, benzcarbazoles and substituted phenols. The benzcarbazole series shown in Figure 5 illustrates two possibly significant factors in carcinogenicity--namely hydrogen bonding and steric (electronic) effects. As expected, the addition of benzene rings increases activity, but the degree of enhancement appears to be strongly dependent on the location of the asymmetric benzene(s) with respect to the pyrrole nitrogen. Thus benzcarbazoles I, II and III with benzene rings adjacent to the pyrrole nitrogen are weak carcinogens, whereas IV, in which the pyrrole NH group is sterically unhindered, is a strong carcinogen. Similarly, while replacement of a benzene ring with pyridine enhances the overall activity of a weak carcinogen (II,III) to a moderate one (V,VI), rotation of the pyridine heterocycle in V through 180° produces a strong carcinogen (VII). In the last structure each N-heterocycle can function independently, since H-bonding is sterically impossible.

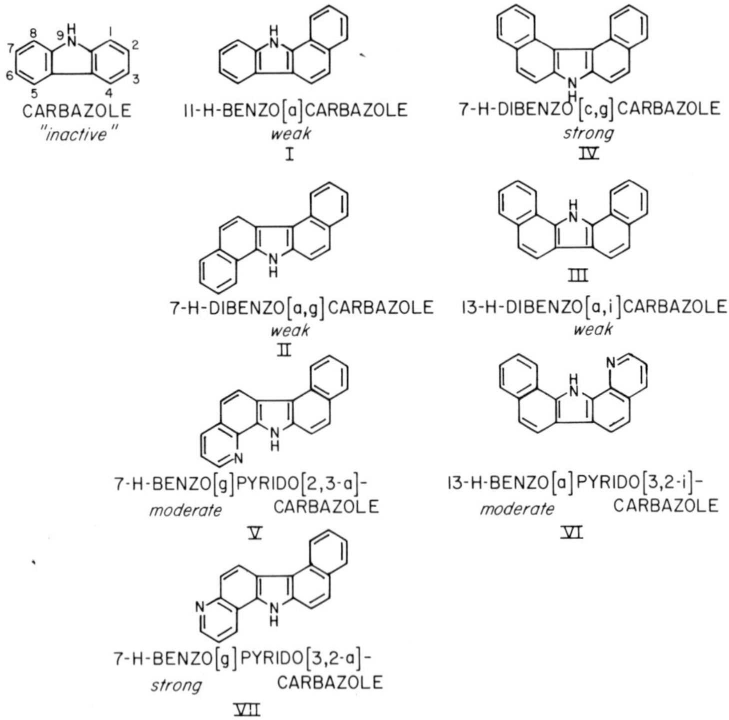

Figure 5. Carcinogenic activities of carbazoles.

MOLECULAR CONNECTIVITY (MC)

Chemists have long known that even minor structural variations in molecules can have profound effects on physical properties, chemical reactivity, and on biological toxicity. This knowledge has been applied to the alteration of structures and substituents of molecules intended for a variety of industrial and medical applications. Two of the most successful applications have been in the areas of drug and polymer design.

Molecular connectivity (MC) attempts to relate these structure-property dependencies to the topological characteristics of the molecule. MC is the most recent and probably the least familiar of the predictive methods used in the correlation of molecular structure with chemical and biological activity. It is based on relatively simple topological principles long familiar to the organic chemist, since a structural formula is in reality a topological graph. Figure 6 illustrates in graph form some simple organic structures with their topological descriptions.

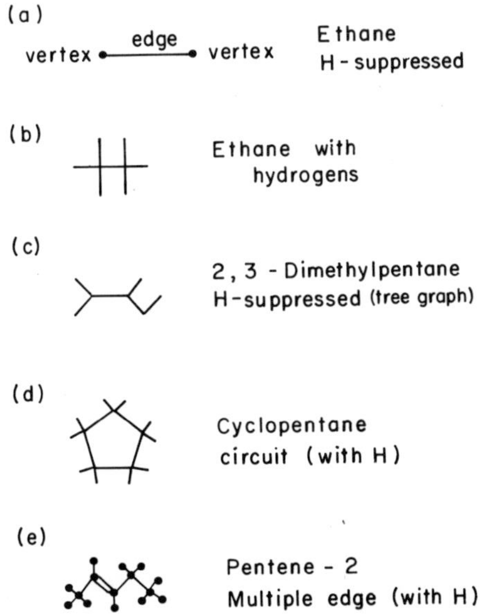

Figure 6. Graph representations of chemical structures.

Molecular connectivity makes the fundamental assumption that it is possible to differentiate molecular structures by abstract numerical means so that their correlation with physical, chemical, and biological properties become possible. The method is defined by Kier as a "non-empirical derivation of numerical values that encode within them sufficient information to relate (them) to many physicochemical and biological properties".[16]

Some Definitions and Simple Graph Theory

A graph is a set of points (vertices) connected by lines (edges; Figure 6). The Molecular Connectivity method assumes that information essential to a quantitative correlation of organic molecular structure with properties is inherent in a valence-weighted graph (G_v). Secondly, a relationship between the connectivity characteristics of the graph and the specified molecular properties is postulated. This relationship is expressed as a sum of terms, each linearly dependent on the graph characteristics.

The connectivity function $C(\chi)$ for a graph has the form

$$C(\chi) = b_o + \sum_{m,t} b_t(m)\,{}^m\chi_t \tag{5}$$

where b_o is a constant and ${}^m\chi_t$ is the connectivity index. Here $b_t(m)$ depends on the property and may be calculated from a model, from theory, or by multiple regression against experimental data. In the latter case, the experimental values are regressed against $C(\chi)$. The number of edges in G_v determines the highest order of the χ term. Each connectivity index term ${}^m\chi_t$ is defined by its subgraph type, t, of m connected edges and subgraph order m. Subgraphs are of the four types listed in Table I. Connectivity Indices are obtained by summing terms over all distinct subgraphs:

$${}^m\chi_t = \sum_{j=1}^{n_m} {}^mS_j \tag{6}$$

where n_m is the number of type t subgraphs of order m. S terms are calculated for each subgraph as reciprocal square root functions of valency,:

$${}^mS_j = \prod_{i=1}^{m+1} (\delta_i)_j^{-1/2} \tag{7}$$

where j refers to a particular set of edges. The number of valencies involved depends on subgraph type. Summation terms ${}^0\chi$ through ${}^4\chi$ are shown in Table II. From these it may be seen that the zero order subgraph consists of a single vertex (no edges); ${}^1\chi$ is summed over all edges, appropriately weighted by reciprocal square root valencies. Here we have only one type of graph edge e_s terminating on v_i and v_j, with a total edge number N_e. Second order subgraphs have pairs of adjacent edges of single path type P. Thus, each term will contain the reciprocal square root product of three vertex valencies.

Table I. Subgraph Types

Order (m)	Valency (δ)	Path Type (t)	Path Descriptor (P)
0	1	---	---
1	2	path	P
2	3	path	P
3	4	path, cluster (star)	P,C
		chain (triangle)	CH
4	5	path, cluster	P,C
		path/cluster	P/C
		chain (cycle)	CH

Table II. Connectivity Indices $^m\chi_t$

Term Order, m	Vertex No., n	Path Type(s)	Equation
0	1	P	$^0\chi = \sum_{i=1}^{n} \delta_i$
1	2	P	$^1\chi = \sum_{s=1}^{N_e} (\delta_i\delta_j)_s^{-1/2}$
2	3	P	$^2\chi = \sum_{s=1}^{n_m} (\delta_i\delta_j\delta_k)_s^{-1/2}$
3	4	P,C,CH	$^3\chi = \sum_{s=1}^{n_m} (\delta_i\delta_j\delta_k\delta_l)_s^{-1/2}$
4	5	P,C,P/C,CH	$^4\chi = \sum_{s=1}^{n_m} (\delta_i\delta_j\delta_k\delta_l\delta_p)_s^{-1/2}$

In the third order connectivity index ($^3\chi$), path, cluster and chain terms may occur, each to be calculated as a reciprocal square root product of four deltas. Finally, in $^4\chi$ all four subgraph types are possible for the first time. Here n_m in the summation term shown refers to the number of type t subgraphs having four edges. $^m\chi_t$ terms of higher order are calculated in a similar manner, with the term superscript m corresponding to the edges involved in the calculation.

Thus, the connectivity index is a valence-weighted count of connected subgraphs. This weighting process is a key feature of the MC method.

Figure 7 illustrates steps in the calculation of the first order connectivity index $^1\chi$ for two isomeric branched aliphatic hydrocarbons (n = 7): 2,2,3-trimethylbutane and 2,4-dimethylpentane. Some useful fundamental equations of MC theory are summarized in Table III.

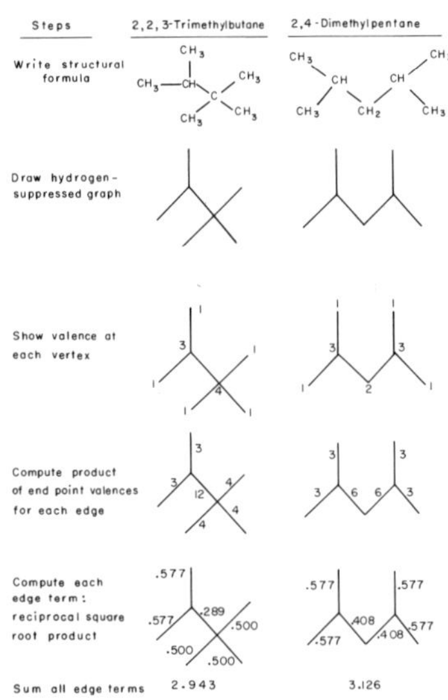

Figure 7. Procedure for calculating connectivity index $^1\chi$.

The topological matrices and alogrithm for dimethylcyclohexane and subgraphs are shown in Table IV. Connectedness values were determined from edge counts E_s:

$$E_s = 1/2 \sum_{i=1}^{n} \sum_{i=1}^{n} A \tag{8}$$

where A is the adjacency matrix.

Table III. Equations of MC Theory

	Equation	Description
(1)	$\sum_{i=1}^{n} \delta i = 2m$	$G(v_i)$, $(i = 1 \ldots n)$ v = vertex = vertex valence m = no. of edges
(2)	$m = n-1$	$G^{m}_{v_n}$ Tree graph
(3)	$\delta i = \sum_{m=1}^{n} T_{ij}$	Vertex valence from summation of row i, matrix T_{ij}
(4)	$C(\chi) = b_o + \sum_{m,t} b\,(m)\,{}^{m}\chi_{t}$	Connectivity function
(5)	${}^{m}\chi_{t} = \sum_{j=1}^{n_m} {}^{m}S_j$	Connectivity index
(6)	${}^{m}S_j = \prod_{i=1}^{m+1} (\delta_i)_j^{-1/2}$	Subgraph term (edge set j, subgraph order m)
(7)	$E_s = 1/2 \sum_{i=1}^{m} \sum_{j=1}^{m} A_{ij}$	Subgraph edges A = subgraph adjacency

Some Uses and Limitations of the MC Method

Using the techniques described, Kier and others have successfully correlated MC with physical and biological properties.[17,18] The method has the advantage of relative simplicity and flexibility. It can be used to represent molecular structure quantitatively at a number of levels of complexity. Each level provides some information uniquely related to the structure (graph, subgraph) and through it to physical, chemical, and biological characteristics.

The $C(\chi)$ function is essentially a weighted count of substructures of the molecule, each described numerically with reference to adjacencies within them. Using similar methods MC may incorporate hetero atoms and valence differences between them. ${}^{1}\chi$ takes account only of adjacent influences on a specific atom. These are modified in the higher order subgraph terms. Also basic to the χ calculation is $\delta_i\delta_j$--the atom product--and use of the reciprocal square roots of this product.

Table IV. Subgraph Evaluation and Enumeration Alogrithm for 1,1-Dimethylcyclohexane[a]

Example Graph	Topological Matrix
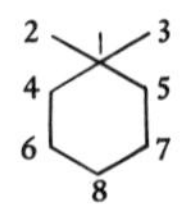	$\begin{pmatrix} 0&1&1&1&1&0&0&0\\ 1&0&0&0&0&0&0&0\\ 1&0&0&0&0&0&0&0\\ 1&0&0&0&0&1&0&0\\ 1&0&0&0&0&0&1&0\\ 0&0&0&1&0&0&0&1\\ 0&0&0&0&1&0&0&1\\ 0&0&0&0&0&1&1&0 \end{pmatrix}$

Vertex Number Sets	Subgraph Matrix	v_i^s	Subgraph	Type
1 2 4 6	$\begin{pmatrix} 0&1&1&0\\ 1&0&0&0\\ 1&0&0&1\\ 0&0&1&0 \end{pmatrix}$	2 1 2 1 $E_s = 3$		Path
1 2 3 4 5	$\begin{pmatrix} 0&1&1&1&1\\ 1&0&0&0&0\\ 1&0&0&0&0\\ 1&0&0&0&0\\ 1&0&0&0&0 \end{pmatrix}$	4 1 1 1 1 $E_s = 4$		Cluster (star)
1 2 3 4 6	$\begin{pmatrix} 0&1&1&1&0\\ 1&0&0&0&0\\ 1&0&0&0&0\\ 1&0&0&0&1\\ 0&0&0&1&0 \end{pmatrix}$	3 1 1 2 1 $E_s = 4$		Path/Cluster
1 4 5 6 7 8	$\begin{pmatrix} 0&1&1&0&0&0\\ 1&0&0&1&0&0\\ 1&0&0&0&1&0\\ 0&1&0&0&0&1\\ 0&0&1&0&0&1\\ 0&0&0&1&1&0 \end{pmatrix}$	2 2 2 2 2 2 $E_s = 6$		Circuit (E_s = subgraph edge count)

[a]Adapted from Ref. 16

Preliminary attempts have been made to develop an atomic chi value $C(^1\bar{\chi}_i)$ in which each bond term (c_{ij}) is divided equally between the two vertices and the half-bond terms summed. Other aspects of molecular structure which require further refinement are: cis-trans isomerism, non-bonded steric interactions and conformational structure, all of which have either three-dimensional or directional features, or both, that are not included in the original treatment. The fact that reasonable correlations have already been achieved for fairly complex systems largely within the limits of an elementary graphical approach is encouraging, for MC is a still maturing technique for coping with the multifactorial problems of chemical-biological interactions.

FACTOR ANALYSIS

The complexity of the chemical mixtures and attendant biological problems associated with oil shale technology clearly preclude the use of a simple SAR or MC approach in the calculation and assessment of environmental and health effects. Thus, while the PAHs contained in the neutral fraction of shale oils have high specific mutagenic activities, they represent only a small fraction of the total activity of a given oil sample because of their low concentrations in shale oils. Consequently the base fraction is viewed as posing the greater health risk. Any predictive method must take these factors into consideration.

One method of dealing with such complex multifactorial data utilizes factor analysis (FA), an approach originally applied mainly to the social sciences and only more recently to chemical and biological problems. Using FA as an analytical tool, it is possible to systematize the multifactorial data of complex interactions without knowledge either of the exact number of significant factors involved or of their precise nature. Thus the assumption is made in QSAR analyses that the biological response (R) of a system to a chemical compound is a function of structure, which in turn has electronic, steric, hydrophobic, and polarizibility terms:

$$R = f(\sigma, E_s, \pi, M_R). \qquad (9)$$

The electronic term is further assumed to be factorable into inductive ($\mathcal{F}$) and resonance ($\mathcal{R}$) terms. Factor analysis provides a rational approach to handling this multiplicity of interrelationships. One need only represent the observed response data in terms of these parameters, or combinations thereof, and develop a model in conformity with the data.

Some Basic Assumptions of FA

A detailed development of FA is beyond the scope of this chapter, and the reader should consult a suitable reference text such as Rummel's Applied Factor Analysis.[19] What follows is a summary of the basic principles of FA applicable to any multivariate system, including our chemical one.

If we assume a two-dimensional data matrix, two mathematical requirements must be satisfied by the property measured. One requirement is that each data point D be expressed as a linear sum of terms:

$$D = d_1 + d_2 + \ldots + d_n. \tag{10}$$

The second requirement is that each data point D also be a sum of row and column product terms:

$$D = r_1c_1 + r_2c_2 + \ldots + r_nc_n \tag{11}$$

where r_k and c_k represent the kth row and column factors, and each is independent of the other. Thus in matrix notation the data matrix may be expressed as the product of a row- and a column-related matrix:

$$[D] = [R].[C]. \tag{12}$$

Matrices may be of the entity-entity or entity-property type. The latter generally has greater relevance in chemical-biological interactive systems. Figure 8 is a diagram of the steps involved in the FA method. The stepwise procedure is described below:

1. Correlation: An experimental data matrix is used to construct a correlation matrix.
2. Decomposition: The correlation matrix is decomposed into a number of linear factors or abstract eigenvectors.
3. Rotation: This is a mathematical operation which relates physically significant parameters to the abstract eigenvectors generated by the preceding operation.
4. Combination: Real factors obtained in step 3 are combined to reproduce a data matrix within the required precision--i.e., to obtain the best solution.
5. Prediction: Good solutions obtained in step 4 are used to predict new data, either by interpolation or by the use of best combinations.

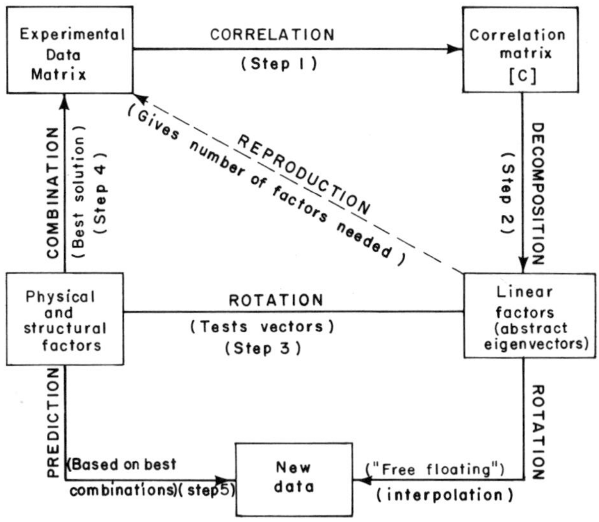

Figure 8. Factor analysis: diagram of the stepwise procedure.

The FA method has been used in a large number of chemical systems and in interactive studies of molecule-biological test pairs.[20-23] It has been used extensively in conjunction with SAR for the study of chemical-biological interactions. The utility of FA and other statistical methods as applied to QSAR has recently been reviewed by Martin.[24]

PATTERN RECOGNITION

Frequently experimental science also requires the prediction of properties not amenable to direct measurement. Thus we may infer elemental composition from atomic emission or absorption spectra; use quantum theory to provide a rational model of atomic structure, and group theory to describe molecular structure. Chemical interactions with bio-systems constitute yet another domain where cause and effect are often obscured by system noise, and their existence must somehow be indirectly inferred from available data. In addition, direct measurement is becoming increasingly less practical for economic and logistic reasons.[25] Here pattern recognition (PR) is a valuable tool for clarifying and ordering information in an efficient and economic manner. It minimizes the quantity of data required and aids in the selection of suitable parameters for achieving the desired correlations.[26]

PR is an expanding branch of artificial intelligence which has long been familiar to biologists, engineers and psychologists.[26,27] The sole assumption made in PR is that a relationship exists between a set of data and a specified category. For a chemical system this category will be defined in terms of, say, a functional group, or some structural feature (e.g., unsaturation, branching). One then attempts to interpret the experimental data obtained in terms of this classification.

Methodology of Pattern Recognition

Operationally the methods of PR fall into two classes, parametric and non-parametric (Figure 9). Parametric methods assume access to probability density functions not usually available for chemical-biological interaction problems. I will therefore confine my remarks to the non-parametric branch of the PR diagram which is devoid of any *a priori* statistical assumptions concerning data distribution.

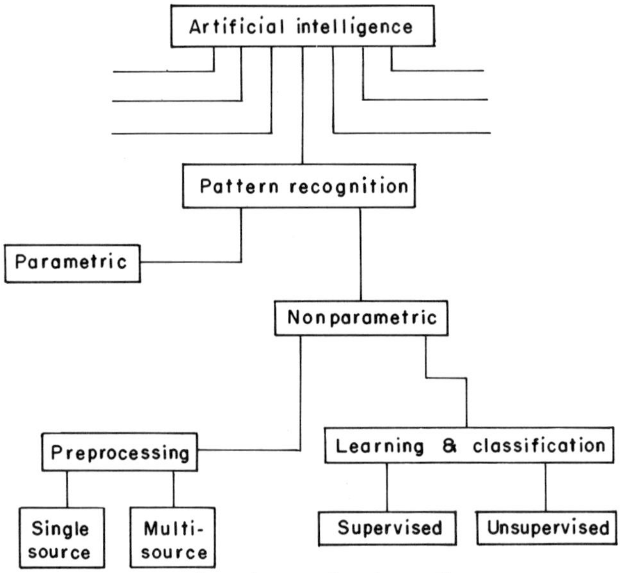

Figure 9. Functional analysis of pattern recognition techniques.

First consider each experimental data point in a collection of measurements as an object in n-dimensional space with coordinates equal to its measurements. The Euclidean distance d_{ij} between any two points, mathematically defined as

$$d_{ij} = [(X_{ik} - X_{jk})^2]^{1/2} \qquad (13)$$

is a measure of their similarity. As similarity between data points i and j increases, the distance between them decreases, approaching zero. Therefore we define a new similarity function S_{ij}, such that

$$S_{ij} = 1 - d_{ij}/D_{ij}. \tag{14}$$

Here D_{ij} is the maximum distance between X_i and X_j and $S_{ij} \to 1$ when d_{ij} assumes a minimum value.

Next, classification and learning processes operate on the n-space in one of two learning modes--supervised or unsupervised. In supervised learning, some of the points are classified and function as a "training set" which can then be used to classify unknown points. In unsupervised learning there is no training set. Instead, the objective is to locate clusters of points in n-space which serve as clues to possibly significant relationships. In either case the basic aim of the PR method is to classify the patterns obtained into well-defined categories.

Preprocessing involves changing the actual structure of points in n-space, and is minimal in the case of unsupervised learning, generally being confined to the scaling of measurements with different units, so as to obtain equal weighting, regardless of the units employed. In the case of supervised learning, data preprocessing may include algebraic transformations, actual changes in variables via mathematical transforms, and feature selection. These operations serve to enhance pattern discrimination by spreading clusters further apart or by reducing the dimensionality of the n-space.

Display of Data: Mapping

If parameters of a system have been judiciously selected with regard to the property being studied, like objects will have similar measurements, hence their proximity in n-space. For n-space > 3 computer techniques can be used to reduce the data to a more manageable 2- or 3-dimensional space. Here the technique of nonlinear mapping (nlm) is often employed to preserve interpoint distances in the ordered space. Figure 10 shows the acid-base separation achieved in a data set abstracted from the periodic table, using the following six properties (n=6) to describe each element: (1) most important valence; (2) melting point; (3) covalent radius; (4) ionic radius; (5) electronegativity and (6) ΔH(fusion). None of these properties used alone could achieve such separation.[28]

Another method of displaying n-dimensional data involves their projection onto a selected 2-dimensional plane after appropriate weighting of the data. This method was used by Ting et al in their successful classification of sixty-six drugs as tranquillizers or sedatives.[29]

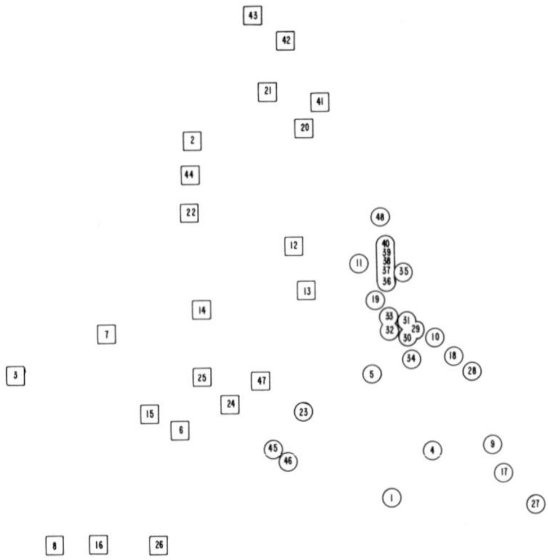

Figure 10. Separation of acids □ and bases ○ for 68 elements of the Periodic Table: nlm from 6-space to 2-space.

Pattern recognition techniques have been applied to the screening of prospective anti-cancer drugs[30] and to structure-activity studies of chemical carcinogens.[31] A detailed discussion of the applications of PR to drug design is given in Ref. 32 by Kirschner and Kowalski.[33] Reports of environmental applications of PR are appearing in the current literature with increasing frequency.[34-37] Extensive references to the application of PR methods to chemical, medical and environmental problems are given in at least three recent books dealing totally or in part with computer assisted methods.[32,38,39]

With the wealth of information already available, can application of PR to the multivariate chemical-biological problems associated with the development of new oil shale and other fossil fuel technologies be far behind?

SUMMARY AND CONCLUSIONS

The potentially toxic byproducts of surface and *in situ* oil shale retorting are too large in number to examine individually. Nor is this desirable, since the insults to humans and ecosystems will be in the form of complex mixtures of toxic gases, particulates, unretorted and spent shales, shale oils and retort waters. Shale oils contain the more volatile toxic trace metals (Hg,As,Sb,Se, etc.) and a large number of organic contaminants which are separable into three complex fractions: neutral, acid and base. Although the neutral fractions contain the most potent carcinogens--the

PAHs--their total amount appears to be low compared with that found in coal and coal-based synfuels, and therefore not of major concern. The base fractions, however, show greater total mutagenic activity and contain a large number of aryl amines and nitrogen heterocyclics which are either known or suspect carcinogens. It is also this fraction which may mobilize the transition metals found in methylene chloride extracts of retort waters by complexing them (Kland, et al).

No single _in vitro_ test applied to such complex mixtures can be expected to be of sufficient universal reliability to provide a criterion for the assessment of potential environmental and health effects. Use of a number of _in vitro_ tests dependent on different biological mechanisms will enhance the reliability of prediction. Best of all is a combination of these with judiciously selected _in vivo_ testing.

The complex nature of both the contaminants and biological test systems involved in the assessment of environmental and health effects of oil shale technologies requires a statistical matrix treatment. Four methods of dealing with such multivariate biological-chemical systems have been described: quantitative structure-activity and molecular connectivity relationships (QSAR,MC), factor analysis (FA) and pattern recognition (PR). QSAR and MC are useful in the prediction of toxic behavior for individual members of a class of compounds for which much SAR data are already available. The QSAR approach uses mathematical functions based on octanol-water distribution coefficients, electronic, steric, hydrophobic and resonance effects, and molar refractions. QSAR is a statistical method. Only objective data are used. It is therefore an excellent predictive tool. However it is not particularly useful in dealing with chemical mixtures, where complex synergistic effects may be operative, and its greatest successes have been achieved in the fields of drug design and the prediction of new drug behavior.

MC utilizes the topological characteristics of molecules to develop nonempirical numerical values based on the connectivities of the atoms within the molecular structure. These are in turn related to actual physicochemical and biological properties of the molecule using a theoretical model or experimental data. The MC approach holds out the very attractive prospect of reducing the essential empirical components of a matrix, and in this respect has some advantage over SAR. It suffers from the same limitations, however, with respect to complex systems. This leaves the methods of factor analysis and pattern recognition, to cope with complex chemical mixtures perturbing biological-environmental systems.

The methods of FA and PR have both been applied to data derived from SARs, and PR techniques have also been used with connectivity functions (Jurs). Thus, both of these methods may be viewed as primarily mathematical techniques for operating on any data to obtain a) more generally applicable solutions in the case of FA and b) reveal clustering or patterns in the data via PR. The biological processes and chemical structures of SARs, while implied, are totally irrelevant. PR is a particularly powerful tool for discrimination--e.g., the detection of relationships in categories of data, *regardless of their significance*. It should most certainly be applied to the data matrix for parameters already measured on Repository samples, with a view to selecting the most useful information, reducing the number of parameters or measurements required, and defining the relationship(s) between processes and data.

PR should also be applied to occupational and health data where currently available from employee and patient profiles (Stallard, this volume, Chapter 6). In the course of a developing synfuels industry much more occupational health data will become available to health professionals. The early application of PR methods to the growing data base could help anticipate problem areas and enable appropriate preventive action before the industry is burdened with costly employee compensation claims and excessive lost-time from job-related causes.

REFERENCES

1. R. O. Beauchamp, Jr. and M. D. Shelby, "Chemicals Identified in Oil Shale and Shale Oil. 1. Preliminary list." Environmental Mutagen Information Center, Information Division, Oak Ridge National Lab., Oak Ridge, TN 37830, 19 pp.

2a. M. J. Kland, H. L. Eaton and A. S. Newton, "Characterization of Trace Contaminants in Oil Shale Retort Waters," Am. Chem. Soc. 35th NW Regional Mtg., Salt Lake City, Utah, June 14, 1980, Paper No. 28. LBL-10850.

b. M. J. Kland, H. L. Eaton, and A. S. Newton, "Trace Contaminants in Oil Shale Retort Waters," in Oil Shale Research: Characterization Studies, 1. E&E Annual Rept. 1979, LBL-10486.

3. M. J. Kland, "The VC-PVC Crisis: A Systematic Approach to Toxicological Problems," Am. Chem. Soc., Symp. on Environ. Chem., 30th NW Regional Mtg., Honolulu, Hawaii, June 12-13 (1975), Paper No. 84. LBL-3275.

4. R. Foster, "Biochemical Systems," Chap. 12 in Charge Transfer Complexes, Academic Press, London (1970), 335-373. See also Chap. 3, "Electronic Spectra," 33-93.

5. Elizabeth C. and James A. Miller, "Metabolism of Chemical Carcinogens to Reactive Electrophiles and Their Possible Mechanisms of Action in Carcinogenesis," Chap. 16 in Chemical Carcinogens, ACS Monograph 173, C. E. Searle, ed., Washington, DC (1976), 737-762.

6a. A. Crum Brown and T. R. Fraser, "On the Connection Between Chemical Constitution and Physiological Action. Part I. (On the) Physiological Action of (the) Salts of (the) Ammonium Bases Derived from Strychnia Brucia, Thebaia, Codeia, Morphia, Nicotia," Trans. Royal Soc. Edinburgh 25, 151-203 (1868).

b. "On the Physiological Action of the Ammonia Bases Derived from Atropia and Conia," XX, Part II. ibid, 693-739 (1869).

7. L. P. Hammett, "Effect of Structure Upon the Reactions of Organic Compounds: Benzene Derivatives," J. Am. Chem. Soc. 59, 96-103 (1937).

8. C. Hansch and T. Fujita, "ρ-σ-π Analysis: (A) Method for (the) Correlation of Biological Activity and Chemical Structure," J. Am. Chem. Soc. 86, 1616-1626 (1964).

9. M. A. H. Fahmy, T. R. Fukuto, R. L. Metcalf and R. L. Holmstead, "Structure-Activity Correlations in DDT Analogs," J. Agric. Fd. Chem. 21(4), 585-591 (1973).

10. R. W. Taft, Jr., "Separation of Polar, Resonance and Steric Effects in Reactivity," in Steric Effects in Organic Chemistry, M. S. Newman, ed., Wiley, New York (1956) 556-675.

11. R. Osman, H. Weinstein and J. P. Green, "Parameters and Methods in Quantitative Structure-Activity Relationships," Chap. 2 in Computer Assisted Drug Design, E. C. Olson and R. E. Christofferson, eds., ACS Symp. Ser. 112, Washington, DC (1979), 21-77.

12. A. Dipple, "Polynuclear Aromatic Hydrocarbons," Chap. 5 in Chemical Carcinogens, Charles E. Searle, ed., ACS Monograph 173 (1976) 245-314.

13. M. Yuan and P. C. Jurs, "Computer-Assisted Structure-Activity Studies of Chemical Carcinogens: A Polycyclic Aromatic Hydrocarbon Data Set," Tox. and Appl. Pharm. 52, 294-312 (1980).

14. D. B. Clayson and R. C. Garner, "Carcinogenic Aromatic Amines," Chapter 8 in Chemical Carcinogens (Ref. 12) 366-461.

15. M. R. Guerin, B. R. Clark, C.-h. Ho, J. L. Epler, and T. K. Rao, "Short-Term Bioassay of Complex Organic Mixtures: Part I. Chemistry," Proc. EPA Symp. on Application of Short-Term Bioassay in the Fractionation and Analysis of Complex Environmental Mixtures, Williamsburg, VA, M. D. Waters, et al., eds., Plenum Press, New York (1978) 247-268.

16. L. B. Kier and L. H. Hall, Molecular Connectivity in Chemistry and Drug Research, Academic Press, New York (1976). Preface.

17. L. B. Kier and W. J. Murray, "Molecular Connectivity, II: Relationships to Biological Activities," J. Med. Chem. 18(12), 1272-1274 (1975).

18. W. J. Murray, L. H. Hall and L. B. Kier, "Molecular Connectivity, III: Relationship to Partition Coefficients," J. Pharm. Sci. 64(12), 1978-1981 (1975).

19. R. J. Rummel, Applied Factor Analysis, Northwestern University Press, Evanston, IL (1970).

20. M. L. Weiner and P. H. Weiner, "A Study of Structure-Activity Relationships of a Series of Diphenylaminopropanols by Factor Analysis," J. Med. Chem. 16(6), 655-661 (1973).

21. R. Wooten, R. Cranfield, G. C. Sheppey and P. J. Goodford, "Physicochemical Activity Relationships in Practice. 2. Rational Selection of Benzenoid Substituents," J. Med. Chem. 18(6), 607-613 (1975).

22. D. G. Howery, "Factor Analyzing the Multifactor Data of Chemistry," Amer. Lab. 8(2), 14-25 (1976).

23. P. H. Weiner and D. G. Howery, "Factor Analysis of Some Chemical and Physical Influences in Gas-Liquid Chromatography," Anal. Chem. 44(7), 1189-1194 (1972).

24. Y. C. Martin, "Other Mathematical Methods of Use in Quantitative Structure-Activity Studies," in Chap. 10 Quantitative Drug Design, Dekker, New York (1978) 233-252.

25. M. J. Kland, "A Priori Predictive Methods of Assessing Health Effects of Chemicals in the Environment," in Water Chlorination: Environmental Impact and Health Effects, Vol. 2, Ann Arbor Science Pub., Ann Arbor, MI (1978) 451-469.

26. T. L. Isenhour and P. C. Jurs, "Some Chemical Applications of Machine Intelligence," in Report for Analytical Chemists, Anal. Chem. 43(10) 20A-35A (1971).

27. J. C. McDonald, "Time-Shared Pattern Recognition," Amer. Lab. 9(2), 31-34 (1977).

28. B. R. Kowalski and C. F. Bender, "Pattern Recognition, A Powerful Approach to Interpreting Chemical Data," J. Am. Chem. Soc. 94(16), 5632-5639 (1972).

29. K-L. H. Ting, R. C. T. Lee, G. W. A. Milne, and A. M. Guarino, "Applications of Artifical Intelligence: Relationships Between Mass Spectra and Pharmacological Activity of Drugs," Science 180 417-420 (1973).

30. B. R. Kowalski and C. F. Bender, "Application of Pattern Recognition to Screening Prospective Anticancer Drugs," J. Am. Chem. Soc. 96, 916-918 (1974).

31. D. S. Dierdorf and B. R. Kowalski, "Three-Dimensional Molecular Structure-Biological Activity Correlations by Pattern Recognition," NTIS Rept. No. AD-785863/2SL (1974), 41 pp.

32. Medicinal Chemistry, VIII, E. J. Ariens, Ed., Academic Press, New York (1979).

33. G. L. Kirschner and B. R. Kowalski, "The Application of Pattern Recognition to Drug Design," Chapter 2 in Ref. 32, 73-131.

34. J. R. McGill and B. R. Kowalski, "Recognizing Patterns in Trace Elements," Appl. Spectroscopy 31, 87-95 (1977).

35. P. K. Hopke, "Application of Multivariate Analysis for Interpretation of the Chemical and Physical Analysis of Lake Sediments," J. Environ. Sci. Health., A11(6), 367-383 (1976).

36. P. D. Gaarenstroom, S. P. Perone and J. L. Moyers, "Application of Pattern Recognition and Factor Analysis for Characterization of Atmosphere Particulate Composition in Southwest Desert Atmosphere," Environ. Sci. Technol. 11(8), 795-800 (1977).

37. P. L. Briggs and F. Press, "Pattern Recognition Applied to Uranium Prospecting," Nature 268, 125-127 (1977).

38. A. J. Stuper, W. E. Brugger and P. C. Jurs, Computer Assisted Studies of Chemical Structure and Biological Function, John Wiley and Sons, New York (1979) 220 pp.

29. Computer-Assisted Drug Design, E. C. Olson and R. E. Christofferson, eds., ACS Symp. Series 112, Washington, DC (1979) 619 pp.

ACKNOWLEDGMENTS

Thanks are due Dr. James C. Bartholomew of the Chemical Biodynamics Division for his review of this paper and helpful comments on the SAR section; to Professor Malcolm M. Renfrew, whose interest in the area of complex environmental systems has resulted in a most stimulating dialog; and to Dr. Christian Daughton of the Shale Group, who reviewed and commented on sections of this chapter.

A. SUBJECT INDEX

B. STUDY MATERIAL INDEX

---

[a]Chemical Repository Sample Number (See Chapter 4).

- JP-8 Product (No. 4609)
  Chem./Phys. 32, 35-37
  Biol. 97-116, 141-172

- Simulated In-Situ Shale Oil (No. 4101)
  Chem./Phys. 163-167
  Biol. 163-167

- Weathered Gasoline Feedstock (No. 4603)
  Chem./Phys. 32, 35-37

Petroleum Derived:

- Mixed Petroleum Crudes (No. 5107)
  Biol. 148-160

- No. 6 Fuel Oil (No. 5401)
  Biol. 189-199

- Petroleum JP-5 (No. 4614)
  Chem./Phys. 32, 35-37, 43
  Biol. 97-116, 129-172, 189-199

- Petroleum JP-8 (No. 4615)
  Chem./Phys. 32, 35
  Biol. 97-116, 141-160

- Petroleum DFM (No. 4616)
  Chem./Phys. 32, 35-37, 42
  Biol. 97-116, 129-172, 189-199

- Recluse Crude Oil (No. 5305)
  Biol. 148-160, 168, 170

- Wilmington Crude Oil (No. 5301)
  Biol. 148-160, 168, 170

Coal Derived:

- COED Syncrude (No. 1106)
  Biol. 157

- H-Coal Distillate (No. 1601), and HDT-Derivatives
  Biol. 158, 170

- H-Coal Fuel Oil, HDT-Derivatives
  Biol. 158

- H-Coal Atmospheric/Vacuum Still Overheads/Bottoms
  Biol. 158